Heating

This fully revised 2nd edition of Keith Moss's highly respected text gives comprehensive coverage of the design of heating and water services in buildings. Each chapter starts with the information needed to understand the specific area, which is then reinforced by many examples and case studies with worked solutions. Mathematics and the principles of fluids are introduced as core skills where they are required as part of the design solution. New material is provided on chimneys, fossil fuel combustion, electrical heating and community and district heating. Students, whether on HNC, HND or degree courses, will find this a valuable book.

Essential reading from Spon Press

Energy Management and Operating Costs
Keith J Moss, UK
ISBN 0–419–21770–3 (pb)

Heat and Mass Transfer in Building Services Design
Keith J Moss, UK
ISBN 0–419–22650–8 (pb)

Building Services Engineering 3rd edition
David Chadderton, ETEQ Pty Ltd, Australia
ISBN 0–419–25730–6 (hb)
ISBN 0–419–25740–3 (pb)

Ventilation of Buildings
Hazim Awbi, Reading University, UK
ISBN 0–419–21080–6 (pb)

**Naturally Ventilated Buildings; buildings for the senses,
the economy and society**
Edited by Derek Clements-Croome, University of Reading, UK
ISBN 0–419–21520–4 (hb)

Information and ordering details
For price availability and ordering visit our website
www.sponpress.com
Alternatively our books and available from all good bookshops.

Heating and Water Services Design in Buildings

2nd edition

Keith J. Moss

Spon Press
Taylor & Francis Group

LONDON AND NEW YORK

First published 2003
by Spon Press
11 New Fetter Lane, London EC4P 4EE

Simultaneously published in the USA and Canada
by Spon Press
29 West 35th Street, New York, NY 10001

Spon Press is an imprint of the Taylor & Francis Group

© 1996, 2003 Keith J. Moss

Typeset in 10/12 Times by
Integra Software Services Pvt. Ltd, Pondicherry, India
Printed and bound in Great Britain by
TJ International, Padstow, Cornwall

British Library Cataloguing in Publication Data
A catalogue record for this book is available
from the British Library

Library of Congress Cataloging in Publication Data
A catalog record for this book has been requested

ISBN 0–415–29185–2 (pbk)
    0–415–29184–4 (hbk)

Contents

Preface

Since 1996 when this book was first published there have been a number of changes that have affected the building services industry. There have been developments in product design as for example, the improvements in variable speed pumps. Less installation work is done on site now and more work completed in the prefabrication workshop.

CIBSE has continued to publish new editions of its Guides. The awarding body Edexcel (BTEC) has published new Programmes for the National Certificate, Higher National Certificate and Higher National Diploma. Engineering Council has made changes to the routes to registration. The new Part L of the Building Regulations is now in force.

The Climate Change Levy has initiated a focus on energy efficient products. Government is now proactive in the matter of sustainable development. For the client, product manufacturer and services engineer, this means accounting for life cycle costing of plant and equipment not only in terms of financial cost but also in terms of costs relating to energy use in extraction of raw materials, transport, product manufacture, operating life and final recycling and waste disposal as well as the cost to the environment.

Clearly this book cannot fully address this issue. However it is a theme that will increasingly influence our industry and the criteria we use for selecting plant and equipment and the design of systems. Where appropriate, reference is made in the text to the new CIBSE Guides. The CIBSE Concise Handbook contains much of the data required in this text and is provided free to student members or available at a very attractive price to non members of the Institution.

Four new chapters have been included in this edition to account for the Heating Units in the new Edexcel Programmes. They are *Flues and draught*, *Combustion of fossil fuels*, *Electric heating* and *District & community heating*.

The text is written in a way that actively involves the reader by encouraging participation in the solutions to examples and case studies, with some examples for the reader to try. Like the first edition it is intended to be a learning text in practical design.

Acknowledgements

I am indebted to Mr Shaw, Mr Sedgley and Mr Douglas, who were my principal teachers at what used to be called the National College in Heating, Ventilating, Air Conditioning and Fan Engineering, and is now integrated with the University of the South Bank.

Grateful thanks are also due to Tony Barton, who preceded me at the City of Bath College and initially set up the courses in HVAC. He it was who introduced a raw recruit from industry to the art of enabling students to learn.

Finally I have to thank all those students who have had to suffer my teaching over the years, because among other things they have taught me that people learn in many different ways, and this makes the profession of teacher a humbling experience and a vocation, in which the teacher is frequently the learner.

Introduction

The profession of building services engineering covers a wide spectrum of the physical sciences. Each part is not an exact science either. Much is learnt on the hoof as it were and so one's initial training needs to be a synthesis of practical experience in the industry and academic learning both of which continue beyond the apprenticeship or trainee period.

This book like the first edition concentrates on the design of 'wet' systems used in commercial and industrial buildings.

Part of Chapter 1 however investigates the way a building behaves when it is intermittently heated, as it is important for you to select the plant and design the systems that will best serve the occupants. It is assumed that you have some knowledge of wet systems otherwise you are directed to current manufacturers' literature so that the text can take on a fuller meaning. If you follow the text through with success and have progressed well with your practical experience in this sector of the industry, you will have attained the equivalent of HND level in heating and water services.

It is strongly recommended that you have access to the CIBSE Guides or the CIBSE Concise Handbook referred to in the Preface, particularly for the pipe-sizing tables. The chapter on steam pipe sizing also requires the use of the Thermodynamic and Transport Properties of Fluids by Mayhew and Rogers. Much of the *underpinning* knowledge in heat transfer and fluid flow is covered in a companion publication.

Heat requirements of buildings in temperate climates **1**

Nomenclature

A	surface area (m^2)
C	specific heat capacity (kJ/kg K)
d	design conditions
dt	temperature difference (K)
dt_t	total temperature difference (K)
f_r	thermal response factor
F_1, F_2	heat loss factors
F_3	plant ratio
H	thermal capacity (kJ/m^2)
K	radiator manufacturer's constant
k	thermal conductivity (W m/m^2 K)
L	thickness of a slab of material (m)
L/k	thermal resistance of the slab (m^2 K/W)
M	mass flow rate (kg/s)
n	space heater index
H	operating plus preheat hours
N	number of air changes per hour
p	prevailing conditions
Q_f	conductive heat loss through the external building fabric (W)
Q_p	plant energy output (continuous heating) (W)
Q_{pb}	boosted plant energy output (intermittent heating) (W)
Q_t	total design heat loss = $Q_f + Q_v$(W)
Q_v	heat loss due to the mass transfer of infiltrating outdoor air (W)
R	fraction of heat radiation
R_a	thermal resistance of the air cavity (m^2 K/W)
R_b	thermal resistance of brick (m^2 K/W)
R_i	thermal resistance of insulation (m^2 K/W)

R_p thermal resistance of plaster (m^2 K/W)
R_{si} inside surface resistance (m^2 K/W)
R_{so} outside surface resistance (m^2 K/W)
R_t total thermal resistance (m^2 K/W)
t_a, t_{ai} indoor air temperature (°C)
t_b balance temperature (°C)
t_c dry resultant, comfort temperature (°C)
t_d datum temperature (°C)
t_{ao} outdoor temperature (°C)
t_f flow temperature (°C)
t_m mean surface temperature (mean radiant temperature) (°C)
t_r return temperature (°C)
t_x temperature of the unheated space (°C)
U thermal transmittance coefficient (W/m^2 K)
V volume (m^3)
VFR   volume flow rate (m^3/s)
Y admittance (W/m^2 K)
ρ density (kg/m^3)
$\sum$ sum of
Cv ventilation conductance (W/K)

1.1 Introduction

Heat flow into or out of a building is primarily dependent upon the prevailing indoor and outdoor temperatures. If both are at the same value, heat flow is zero, and the indoor and outdoor climates are in balance with no heating required.

During the heating season (autumn, winter and spring), when outdoor temperature can be low, the space heating system is used to raise the indoor temperature artificially to a comfortable level, resulting in heat losses through the building envelope to the outdoors. The rate of heat loss from the building depends upon:

- the heat flow into or through the building structure, Q_f (in Watts);
- the rate of infiltration of outdoor air, resulting in heat flow Q_v (in Watts) to the outdoors as the warmed air exfiltrates;
- building shape and orientation;
- geographical location and exposure.

A single storey building will have a greater heat loss than a multi-storey building of the same floor area since there is little or no heat loss through the upper floors. On the other hand, the upper floors of a multi-storey building are more exposed to the weather.

STRUCTURAL HEAT LOSS

This occurs as heat conduction at right angles to the surface, and is initially expressed as *total thermal resistance* R_t, where

$$R_t = R_{si} + \sum \frac{L}{k} + R_a + R_{so} \quad (m^2\,K/W) \qquad (1.1)$$

The *thermal transmittance coefficient*, $U = 1/R_t$ (W/m²K), is the rate of conductive heat flow through a composite structure (consisting of a number of slabs of material, which can include air cavities) per square metre of surface, and for one degree difference between the indoor and outdoor temperature.

It follows that structural heat loss from the exposed building envelope is given by

Fabric Heat Loss

$$Q_f = \sum (UA)dt \quad (W) \qquad (1.2)$$

INFILTRATION

Consider Figure 1.1, which shows a section through a building. The prevailing wind infiltrates one side of the building, where the space heating appliances must be sized to raise the temperature of the incoming outdoor air. The heated air moves across the building to the leeward side, where it is exfiltrated. Here the heating appliances are not required to treat the air. Clearly, on another day the wind direction will have changed, and therefore all heating appliances

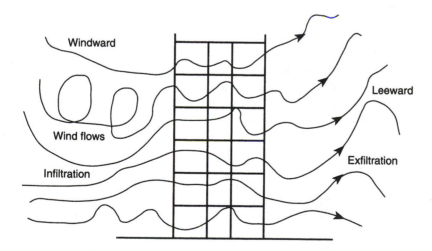

Figure 1.1   Section through a multi-storey building showing prevailing wind pattern.

serving the building perimeter must be sized to offset the infiltration loss Q_v as well as the structural heat loss Q_f.

Theoretically, the building heat loss Q_t would therefore be the sum of the total structural heat loss plus only half of the total loss due to infiltration, as only half of the appliances are exposed to cold infiltrating air at any instant. In practice, the full infiltration heat loss is included in the building heat loss calculation.

It follows that the rate of infiltration – and hence the heat loss due to infiltration of outdoor air Q_f is dependent upon air temperature and wind speed. Further factors that relate to the building design will also influence the infiltration rates, and include stack effect in the building resulting from stairwells, lift shafts, unsealed service shafts and atria, and how well the building is sealed.

Thus from $Q = MC \, dt$, $Q_v = (V \rho NC \, dt)/3600$, and if standard values of air density and specific heat capacity are taken, the constant will be: $\rho C/3600 = 1.2 \times 1010/3600 = 0.33$. Thus $Q_v = 0.33NV \, dt$. Since $Cv = NV/3$,

$$Q_v = Cv \, dt \quad \text{(W)} \tag{1.3}$$

BUILDING SHAPE AND ORIENTATION

This affects the way indoor temperature control is achieved to offset the effects of solar irradiation on and through the building envelope (see Chapter 6).

GEOGRAPHICAL LOCATION AND EXPOSURE

The location and elevation of the site are accounted for in the choice of outdoor design temperature. In the absence of local knowledge, information and data are offered in the *CIBSE Guide A*.

Exposure relates to the effect of wind speed, the increase of which gradually destroys the outside surface resistance R_{so}, which increases the thermal transmittance U and increases the rate of infiltration N.

BUILDING HEAT LOSS

Irrespective of the *type* of space heating system being used in a building, the total heat loss at design conditions will be

$$Q_t = Q_f + Q_v \quad \text{(W)}$$

This may be written as $Q_t = \left(\sum (AU) + Cv \right) (t_c - t_{ao})$.

1.2 Heat energy flows

There are two generic observations that apply to the natural world:

- Heat energy will always flow from a high-temperature zone to zones at lower temperature.
- The rate of heat flow is dependent upon the magnitude of the temperature difference between zones.

For example, if the heat flow from a building is 100 kW when t_c is 20 °C and t_o is −3 °C, where the magnitude of the temperature difference is 23 K, it is clear that heat flow will increase when the outdoor temperature falls to −10 °C and the temperature difference is now 30 K. The revised heat loss = 100 × 30/23 = 130.4 kW.

Likewise the output of a radiator varies with the magnitude of the temperature difference between its mean surface temperature t_m and room temperature t_c. Consider Figure 1.2, which shows a section through a building and the heat flow paths expected during the winter season. When temperatures are steady, a heat balance may be drawn:

heat loss from the building = heat output from the space heater

Using appropriate equations:

$$\left(\sum(UA) + Cv\right)(t_c - t_{ao}) = KA(t_m - t_c)^n \qquad (1.4)$$

Index n is approximately 1.3 for radiators and 1.5 for natural draught convectors, and is found empirically.

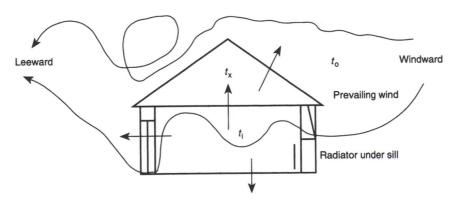

Figure 1.2 Section through building: $t_i > t_x > t_o$.

Example 1.1
(a)   A room has a heat loss of 6 kW when held at a temperature of 20 °C for an outdoor temperature of −1 °C. Find the required surface area of a radiator to offset the room heat loss, given that the manufacturer's constant $K = 13\,\text{W/m}^2\,\text{K}$, index $n = 1.3$ and the flow and return temperatures at the radiator are 80 °C and 70 °C respectively.
(b)   If the outdoor temperature rises to 5 °C, find the required mean surface temperature of the radiator to maintain the room at a constant 20 °C.

Solution
(a)   mean temperature $t_m = 0.5(80 + 70) = 75\,°\text{C}$. Adopting equation (1.4),

$$6000 = 13A(75 - 20)^{1.3}$$

from which $A = 2.52\,\text{m}^2$.
(b)   Prevailing heat loss is the product of design loss and the ratio of temperature differences:

$$Q = 6000\left(\frac{20 - 5}{20 + 1}\right) = 4286\,\text{W}$$

Substituting into equation (1.4),

$$4286 = 13 \times 2.522(t_m - 20)^{1.3}$$

from which

$$t_m = 62\,°\text{C}$$

You should now confirm this solution. These results are summarized in Table 1.1.

The system controls must vary the radiator mean surface temperature as the outdoor temperature varies. However, the *rate* of response required for changes in outdoor climate is dependent upon the thermal capacity of the building envelope, and this varies from lightweight structures, which have a short thermal response, to heavyweight structures.

Table 1.1   Example 1.1

Conditions	Heat loss (W)	Radiator output (W)	$(t_m - t_i)$ (K)	Outdoor temperature (°C)
Design	6000	6000	55	−1
Prevailing	4286	4286	42	+5

Example 1.2

A natural draught convector circuit has design conditions $t_c = 20\,°C$, $t_o = -1\,°C$, $t_f = 82\,°C$ and $t_r = 70\,°C$. Determine the required mean water temperature and the circuit flow and return temperatures to maintain a constant indoor temperature when outdoor temperature rises to $7\,°C$. Take index n as 1.5.

Solution

Here the energy balance may be extended (equation 1.4), if it is again assumed that temperatures remain steady:

heat loss = convector output = heat given up by the heating medium

The last part of the heat balance is obtained from

$$Q = MC(t_f - t_r) \quad (\text{W})$$

assuming the heating medium is water.

Under operating conditions the constants U, A, Cv, K, A, M and C in each part of the heat balance can be ignored, and

$$(t_c - t_{ao}) \propto (t_m - t_c)^n \propto (t_f - t_r)$$

If design (d) and prevailing (p) temperatures are put together:

$$\frac{(t_c - t_{ao})_p}{(t_c - t_{ao})_d} = \left(\frac{(t_m - t_c)_p}{(t_m - t_i)_d}\right)^n = \frac{(t_f - t_r)}{(t_f - t_r)_d} \qquad (1.5)$$

Equating heat loss with heat output,

$$\frac{20 - 7}{20 + 1} = \frac{(t_m - 20)^{1.5}}{(76 - 20)^{1.5}}$$

from which

$$t_m = 60.7\,°C$$

You should now confirm that this is so.

Equating heat loss with heat given up,

$$\frac{13}{21} = \frac{dt}{82 - 70}$$

Table 1.2   Example 1.2

				Condition				
t_i	t_o	dt	t_f	t_r	dt	t_m	$(t_m - t_i)$	dt
Design	20	−1	21	82	70	12	76	(76 − 20)	56
Prevailing	20	+7	13	64.4	57	7.4	60.7	(60.7 − 20)	40.7

from which

$$dt = 7.4\,\text{K}$$

For two-pipe distribution:

$$t_f = t_m + 0.5\,dt = 60.7 + 3.7 = 64.4\,°\text{C}$$

and

$$t_r = t_m - 0.5\,dt = 60.7 - 3.7 = 57\,°\text{C}$$

The results are summarized in Table 1.2.

Conclusion

A series of flow temperatures may be evaluated for corresponding differ-ent values of outdoor temperatures, and a plot of outdoor temperature versus circuit flow temperature has been produced. This provides the basis for calibrating the controls. See Figure 1.3.

At an outdoor temperature of 15 °C it is assumed here that there are sufficient indoor heat gains to keep the indoor temperature at design at 20 °C without the use of the space heating system. If this is the case, the *balance temperature* t_b is 15 °C.

The balance temperature can be calculated from

$$t_b = t_i - \left(\frac{\text{heat gain (W)}}{\text{design heat loss (W/K)}} \right)$$

Circuit flow temperature of 30 °C is an arbitrary value, and depends upon calibration.

It is worth noting again that a change in outdoor condition will not require an immediate response from the controls to maintain a constant indoor temperature. The response time will depend upon the thermal capacity of the building envelope.

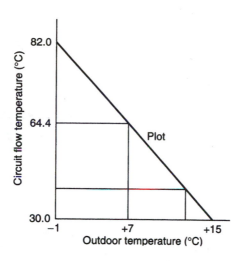

Figure 1.3 Calibration of temperature controls.

THERMAL CAPACITY OF THE BUILDING ENVELOPE

The thermal capacity H of the building envelope is normally measured in kJ/m^2 of structural surface on the warm side of the insulation slab, which may consist of a proprietary material, or it may have to be taken as the air cavity if there is no identifiable thermal insulation slab within the structure.

When the space heating plant starts up after a shutdown period of, say, a weekend, the building envelope is cold, and heat energy is absorbed into the structural layers on the room side of the insulation slab until optimum temperatures are reached in the layers of material. At this point, the rooms should begin to feel sufficiently comfortable to occupy. The more layers of material there are on the room side before the insulation slab is reached, the greater will be the thermal capacity of the building envelope and the longer the preheat period. Conversely, the longer is the cool-down period after the plant is shut down.

The energy equation is

$$H = \text{slab thickness } L \times \rho \times C \times (t_m - t_d) \quad (\text{kJ/m}^2) \qquad (1.6)$$

Example 1.3
Consider the composite walls (a) and (b) detailed in Figure 1.4. From the data, determine the wall thermal capacity on the hot side of the insulation slab for each case and draw conclusions from the solutions.

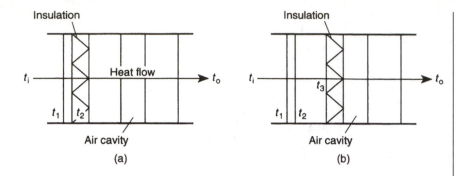

Figure 1.4 Two similar walls with insulation in different locations.

Data

Wall elements are: 10 mm lightweight plaster, $k = 0.16 \, \text{W/mK}$; 25 mm mineral fibre slab, $k = 0.035 \, \text{W/mK}$; 100 mm brick, $k = 0.62 \, \text{W/mK}$; air cavity, $R_a = 0.18 \, \text{m}^2 \, \text{K/W}$; 100 mm brick, $k = 0.84 \, \text{W/mK}$. Surface resistances $R_{si} = 0.12 \, \text{m}^2 \, \text{K/W}$, $R_{so} = 0.06 \, \text{m}^2 \, \text{K/W}$. Plaster density $600 \, \text{kg/m}^3$, specific heat capacity $1 \, \text{kJ/kg K}$; brick density $1700 \, \text{kg/m}^3$, specific heat capacity $0.8 \, \text{kJ/kg K}$.

Note: (i) that the thermal insulation slab is located differently in each case. Indoor temperature $20 \, °\text{C}$, outdoor temperature $-1 \, °\text{C}$ and datum temperature is taken as $12 \, °\text{C}$; (ii) that the outer leaf of the brick wall has a higher thermal conductivity since it is subject to rain penetration.

Solution

For wall (a) given that $R = L/k \, (\text{m}^2 \, \text{K/W})$:

$$R_t = R_{si} + R_p + R_i + R_b + R_a + R_b + R_{so}$$
$$= 0.12 + 0.0625 + 0.7143 + 0.1613 + 0.18 + 0.119 + 0.06$$
$$= 1.4171 \, \text{m}^2 \, \text{K/W}$$

As thermal resistance $R \propto \mathrm{d}t$:

$$\frac{R}{R_t} = \frac{\mathrm{d}t}{\mathrm{d}t_t} \tag{1.7}$$

Thus

$$\frac{R_{si}}{R_t} = \frac{t_i - t_1}{t_i - t_o}$$

Substituting

$$\frac{0.12}{1.4171} = \frac{20 - t_1}{20 + 1}$$

from which

$$t_1 = 18.22\,°C$$

Similarly

$$\frac{0.12 + 0.0625}{1.4171} = \frac{20 - t_2}{20 + 1}$$

from which

$$t_2 = 17.3\,°C$$

The mean temperature of the plaster, t_m, is therefore given by

$$t_m = \frac{18.22 + 17.3}{2} = 17.76\,°C$$

and the thermal capacity of the plaster, which is the only element on the hot side of the insulation, can now be determined from equation (1.6) and knowledge of the density and specific heat capacity of the lightweight plaster. Thus

$$H = 0.01 \times 600 \times 1.0(17.76 - 12)$$

from which, for wall (a)

$$H = 34.6\,kJ/m^2$$

For wall (b), the total thermal resistance R_t remains the same, but the slabs of material are arranged so that now the material on the warm side of the insulation includes the plaster and the inner leaf of the wall.

The inside surface temperature t_1 and temperature t_2 at the interface will have the same value. Interface temperature t_3 needs calculation, and from equation (1.7),

$$\frac{R_{si} + R_p + R_b}{R_t} = \frac{t_i - t_3}{t_i - t_o}$$

Substituting

$$\frac{0.12 + 0.0626 + 0.1613}{1.4171} = \frac{20 - t_3}{20 + 1}$$

from which

$$t_3 = 14.9\,°C$$

The mean temperature of the inner brick leaf of the wall therefore will be

$$t_m = \frac{17.3 + 14.9}{2} = 16.1\,°C$$

The thermal capacity on the hot side of the insulation now includes the plaster plus the inner brick leaf, for which density and specific heat capacity are required, and

$$H = 33.6 + (0.1 \times 1700 \times 0.8)(16.1 - 12)$$
$$= 34.6 + 557.6$$
$$= 592\,kJ/m^2$$

The datum temperature t_m of $12\,°C$ is, in this context, the point above which heat energy is measured. The indoor frost thermostat for temperature control to a building might be set at $12\,°C$, this being the temperature below which it is not desirable for the building envelope to go when the plant is normally inoperative. This therefore seems a reasonable datum temperature to adopt.

Conclusion

The effect on the thermal capacity in wall (b) is considerable. This implies that before comfort levels are reached, the external wall will need to absorb $592\,kJ/m^2$ of heat energy from the space heating plant. The following analyses may be made.

1. Slab density has a significant effect on thermal capacity on the hot side of the thermal insulation slab.
2. It will take longer for comfort conditions to be reached in wall (b) than in wall (a). Thus the preheating period will need to be longer. Conversely, cooling will take longer, allowing the plant to be shut down earlier.
3. It will take longer for the inside surface temperature of $18.22\,°C$ to be reached in wall (b).
4. The thermal transmittance coefficient (U) is the *same* for both wall (a) and wall (b).

5. The thermal admittance (Y) is *lower* in wall (a) than in wall (b).
6. The location of the thermal insulation slab in the structure dictates the *thermal response factor* (f_r) for that structure. It can effectively alter a sluggish thermal response (heavyweight) structure to a rapid thermal response (lightweight) structure when the insulation slab is located at or near the inside surface.
7. The inner leaf of the wall in composite wall (b) should consist of lightweight block if a faster response is required.
8. The air cavity may be taken as the insulation slab in the absence of insulation material in a composite external structure when identifying the slabs on the hot side.
9. If insulated lining is added on the inside of an external wall during refurbishment, the inside surface temperature of the wall is raised, thus improving comfort; the U and Y values are each reduced, thus saving energy; preheat and cool-down times are reduced; and the wall is behaving like a lightweight structure.
10. Internal walls and intermediate floors, which are not exposed to the outdoor climate, and which are constructed from dense materials like concrete and block-work, will have a flywheel or damping effect upon the rise and fall in temperature of the building structure for an intermittently operated plant. They in effect act as a heat store.

VAPOUR FLOW

Air at an external design condition of $3\,°C$ during precipitation (rainfall) can be at saturated conditions (relative humidity of 100%). If it is then sensibly heated to $20\,°C$ dry bulb by passing it through an air heater battery, its relative humidity drops to 32%. At both dry bulb temperatures the vapour pressure remains constant at $7.6\,mbar$. This is because moisture in the form of latent heat has not been absorbed from or released to the air.

The partial pressure of the water vapour in the air is therefore altered only by adding or removing latent heat through the process of evaporation or condensation. Indoor latent heat gains are incurred immediately upon the building being occupied, owing to involuntary evaporation from the skin surface, exhalation of water vapour from the lungs and sweating. In the winter, therefore, latent heat gain indoors is inevitable during occupancy periods. Cooking, dishwashing and laundering add to the latent heat gains. If the building is heated, the air is usually able to absorb the vapour production, with a consequent rise in vapour pressure. In an unheated and occupied building, condensation may occur on the inside surface of the external structure because the air is unable to absorb all the vapour being produced.

Vapour pressure in heated and occupied buildings is inevitably higher than the vapour pressure in outdoor air. Vapour therefore will migrate from indoors to outdoors. In ventilated buildings this may well take place via the ventilating air. Otherwise water vapour will migrate through the porous elements of the building envelope. If the temperature gradient in the external structure reaches dew-point, the migrating vapour will condense. It is important to ensure that, when it does, it occurs in the external leaf of the structure, which is usually capable of saturation from driving rain. The use of vapour barriers in the building envelope is also common practice. It is important to ensure that the vapour barrier is located as near to the warm side (i.e. the indoors) of the external structure as possible. It is also important to know that vapour barriers only inhibit the migration of water vapour unless materials like glass, plastic or metals are used, and even here migration can occur around seals and through the smallest puncture. However, if vapour migration is largely inhibited, the likelihood of interstitial condensation (that occurring within the external structure) is rare.

The occurrence of condensation and dampness on the inside surface of the building envelope is avoided only by adequate ventilation and thermal insulation. There are software programs that can identify the incidence of surface and interstitial condensation for given external composite structures with and without the use of the vapour barrier.

1.3   Plant energy output

The determination of building heat loss Q_t, which forms the basis for the calculation of plant energy output is dependent upon the mode of plant operation and hence the mode of occupancy. It can be divided into three categories: continuous, intermittent and highly intermittent.

It has long been known that building envelopes for factories and workshops having traditional transmittance coefficients (average U value approximately $1.5\,\text{W/m}^2\,\text{K}$) and relatively high rates of infiltration (above 1 air change/h) require higher levels of convective heating than radiant heating to maintain a comfortable environment. This is based on the knowledge that, living in the natural world as we do, we are quite comfortable in outdoor climates of relatively low air temperature and velocity if solar radiation is present with sufficient intensity. A similar response is to be found indoors when a significant proportion of appliance heat output is in the form of heat radiation.

The calculation of building heat loss Q_t using heat loss factors F_1 and F_2 accounts for the varying proportions of radiant and convective heating offered by different heating appliances in such building envelopes. However, it will be shown that for buildings subject to current thermal insulation standards (average U value 0.5 or less) and with infiltration rates below 1/h (i.e. substantially airtight), the heat loss has about the same value for both highly radiant and highly convective systems.

The decision regarding which type of space heating system to select for modern factories and workshops can now be centred on the use to which the building will be put. For example, high-tech dust-free environments will generally benefit from radiant systems with very low air movement. Indoor environments in which dust is not such a problem can be heated using an air-heating system.

A good measure of thermal comfort can be found from the closeness of three thermal indices namely air, dry resultant and mean radiant temperature. Dry resultant or comfort temperature is the indoor design index. Air and mean radiant temperatures can be found from the following equations:

$$t_{ai} = \frac{Q_t(1 - 1.5R) + Cvt_{ao} + 6\sum At_c}{Cv + 6\sum A} \qquad (1.8)$$

when air velocity approaches 0.1 m/s,

$$t_c = 0.5t_m + 0.5t_{ai} \quad \text{so} \quad t_m = 2t_c - t_{ai} \qquad (1.9)$$

The heat loss factors F_1 and F_2 can be found from

$$F_1 = \frac{3(Cv + 6\sum A)}{\sum(AU) + 18\sum A + 1.5R(3Cv - \sum(AU))} \qquad (1.10)$$

$$F_2 = \frac{\sum(AU) + 18\sum A}{\sum(AU) + 18\sum A + 1.5R(3Cv - \sum(AU))} \qquad (1.11)$$

Factor R is a variable, and primarily depends upon the proportion of radiant to convective heat output from the space heating system. Table 1.3 gives the values of R.

Table 1.3   Values of factor R

Heat radiation proportion (%)	R
0	0
10	0.10
20	0.20
30	0.30
50	0.40
67	0.67
90	0.90

The steady state model of total design heat loss is obtained from

$$Q_t = \left(F_1 \sum (UA) + CvF_2 \right)(t_c - t_{ao})$$ (1.12)

This equation is effectively the sum of the fabric heat loss Q_f and the heat loss due to natural infiltration, Q_v. Q_f and Q_v can be determined separately if required.

CONTINUOUS HEATING

Plant energy output Q_p for a continuously heated building classified as heavyweight can equal the design building heat loss Q_t. The building envelope acts as a heat store and because of its high thermal inertia will not react to short periods of severe weather. Buildings classified as lightweight on the other hand tend to be sensitive to outdoor temperature change.

In practice, it is normal to provide a plant ratio F_3 for sizing a boiler plant that operates continuously during the heating season. This will vary from around 1.1 for buildings classified as heavyweight to 1.25 for lightweight buildings. Thus

$$Q_p = F_3 \times Q_t$$

where F_3 will be between 1.1 and 1.25. There now follows the routine for calculating the design heat loss Q_t and plant energy output Q_p.

Example 1.4
Determine the design heat loss Q_t and Q_p for a fully exposed workshop measuring $30 \times 15 \times 5\,\mathrm{m}$ high having a 'flat' roof, where the anticipated air change rate is 2.5/h:

(a) for a system of unit heaters (100% convective);
(b) for a system of gas-heated radiant tubes (10% convective, 90% radiant).

Data
$t_c = 18\,°\mathrm{C}$, $t_{ao} = -2\,°\mathrm{C}$, area of wall glass $= 135\,\mathrm{m}^2$. Thermal transmittance coefficients: $U_g = 5.7$, $U_w = 1.3$, $U_r = 1.5$ and $U_f = 1.0\,\mathrm{W/m^2\,K}$ (the suffixes refer to glass, wall, roof and floor respectively). Plant ratio F_3 is estimated at 1.25.

Solution
(a) The structural heat loss is tabulated in Table 1.4. From equation (1.10):
$$F_1 = 1.125$$

Table 1.4 Example 1.4: tabulation of structure heat loss

Element	Dimensions (m)	Area, A	U	UA
Wall glass		135	5.7	769.5
Wall	$(90 \times 5) - 135$	315	1.3	409.5
Roof	30×15	450	1.5	675
Floor	30×15	450	1.0	450
		$\sum A$ 1350		$\sum (UA)$ 2304

From equation (1.11):

$$F_2 = 1.0$$

You should now confirm the values of F_1 and F_2. From equation (1.12) and Table 1.4:

$$Q_t = ((1.125 \times 2304) + (1.0 \times 1875))(18 + 2) = 89\,340 \text{ W}$$

$$Q_p = 89\,340 \times 1.25 = 111\,675 \text{ W}$$

(b)   From equation (1.10):

$$F_1 = 0.963$$

From equation (1.11):

$$F_2 = 0.856$$

You should now confirm the values for the heat loss factors. From equation (1.12) and Table 1.4:

$$Q_t = ((0.963 \times 2304) + (0.856 \times 1875))(18 + 2) = 76\,480 \text{ W}$$

$$Q_p = 76\,480 \times 1.25 = 95\,600 \text{ W}$$

A comparison of the results shows an increase of 17% in heat loss for the system of unit heaters over the radiant system. However, a comparison of temperatures is also revealing.

The solutions are tabulated for analysis in Table 1.5. Note the variations in the indoor thermal indices in both cases. For a thermally comfortable environment these indices should be approaching comfort temperature t_c. Note that the fabric and ventilation losses have been calculated separately using equation (1.12). The heat flow path from indoors to outdoors is different in each case and identifies from where the heat energy originates, namely from the air, t_{ai}, in (a) and from the surface (t_m being the mean value of the surfaces) in (b), as shown in Figure 1.5.

(a)   Unit heaters

From equation (1.8):

$$t_{ai} = \frac{89\,340 - 3750 + 145\,800}{1875 + 8100}$$

from which

$$t_{ai} = 23.2\,°C$$

From equation (1.9)

$$t_m = (2 \times 18) - 23.2$$

from which

$$t_m = 12.8\,°C$$

(b)   Gas fired radiant tubes

From equation (1.8):

$$t_{ai} = \frac{-26\,768 - 3750 + 145\,800}{1875 + 8100}$$

from which

$$t_{ai} = 11.6\,°C$$

From equation (1.9)

$$t_m = (2 \times 18) - 11.6$$

from which

$$t_m = 24.4\,°C$$

Table 1.5   Example 1.4: comparison of temperatures

System	Q_t (kW)	Q_f (kW)	Q_v (kW)	t_c (°C)	t_{ai} (°C)	t_m (°C)	t_{ao} (°C)
Warm air	89.3	51.8	37.5	18	23.2	12.8	−2
90% radiant	76.5	44.4	32.1	18	11.6	24.4	−2

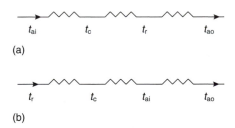

(a)

(b)

Figure 1.5   Heat flow paths for (a) convective and (b) radiant heating.

Example 1.5
If the same workshop is now considered to have an envelope constructed to current thermal insulation standards a further calculation and analysis may be undertaken. Revised thermal transmittance coefficients: $U_g = 3.0$, $U_w = 0.6$, $U_r = 0.25$ and $U_f = 0.3\,W/m^2\,K$. Air change rate $N = 0.75/h$. Plant ratio F_3 is estimated at 1.1.

Solution
(a)   Unit heaters
From equation (1.10):

$$F_1 = 1.034$$

From equation (1.11):

$$F_2 = 1.0$$

From equation (1.12):

$$Q_t = 28\,652\,\text{W}$$
$$Q_p = 28\,652 \times 1.1 = 31\,517\,\text{W}$$

(b)   Radiant heaters
From equation (1.10):

$$F_1 = 0.989$$

From equation (1.11):

$$F_2 = 0.956$$

From equation (1.12):

$$Q_t = 27\,400\,\text{W}$$
$$Q_p = 27\,400 \times 1.1 = 30\,140\,\text{W}$$

Clearly the difference between the two heat loss totals is insignificant.

Comparison of temperatures
For unit heaters, from equation (1.8):

$$t_{ai} = \frac{28\,652 - 1125 + 145\,800}{8662.5}$$

from which

$$t_{ai} = 20\,°\text{C}$$

From equation (1.9):

$$t_m = (2 \times 18) - 20$$

from which

$$t_m = 16\,°\text{C}$$

Table 1.6 Example 1.5: solutions tabulated for analysis

System	Q_t (kW)	Q_f (kW)	Q_v (kW)	t_c (°C)	t_{ai} (°C)	t_m (°C)	t_{ao} (°C)
Warm air	28.7	17.4	11.3	18	20	16	−2
Radiant	27.4	16.6	10.8	18	15.6	20.4	−2

The temperature interrelationships in (b) for the radiant system are calculated and found to be, $t_{ai} = 15.6\,°C$ and $t_m = 20.4\,°C$.

The solutions are tabulated for analysis in Table 1.6. Note that for both warm air and radiant heating the thermal indices are closer together, so comfort should be achieved with either system. The total value for Q_t in each case is very close, indicating that the proportions of convective and radiant heating are not important factors for a well-insulated building with low infiltration.

Conclusion

(i)   The conclusion that may be drawn from these analyses is that for well-insulated buildings with air change rates less than 1/h, the plant energy output is approximately the same for any system of heating regardless of the proportions of radiant to convective output.

However, for poorly insulated buildings and buildings with high air change rates, the type of space heating assumes importance in the determination of building design heat loss Q_t.

(ii)   The calculation of plant energy output Q_p for a continuously heated building will be $Q_p = F_3 Q_t$ where plant ratio F_3 will lie between 1.1 and 1.25.

INTERMITTENT HEATING

Examples 1.4 and 1.5 are based on the assumption that the heating plant operates continuously throughout the heating season. This is the case for factories working a system of three shifts, hospitals, airport terminals, residential homes, etc. However, there are many examples where plants operate intermittently, and this is likely to include night time and weekend shutdown.

Clearly equation (1.12) must be adjusted, as an additional capacity will be required of generating plant to provide a boosted output Q_{pb} during the preheat period before occupation. Thus

$$Q_{pb} = F_3 \times Q_t \tag{1.13}$$

F_3 is the *plant ratio*, and for intermittent heating, it must be calculated. This is determined from

$$F_3 = \frac{24f_r}{Hf_r + (24 - H)} \qquad (1.14)$$

The *thermal response factor* (f_r) takes into account the thermal inertia of the building fabric. The greater the thermal storage capacity of the structure, the higher will be the thermal inertia and the greater will be its mass. Thermal response is dependent upon the heat flow *into* the structure as well as the heat flow *through* the structure. These two characteristics are identified as *absorption* (admittance Y) and *transmission* (transmittance U). The admittance Y is determined from heat flow into the immediate layers on the inside of the structure. Its value is dependent upon the density and thermal storage capacity of these layers of fabric on the room side of the building envelope.

Admittance is not therefore easily calculated. Values of Y for different composite structures are given alongside the transmittance (U) values in section A of the *CIBSE Guide*. As heating plant becomes more intermittent in operation, building heat loss becomes increasingly sensitive to heat flow *into* rather than heat flow *through* the fabric envelope.

Thermal response is determined from

$$f_r = \frac{\sum(YA) + Cv}{\sum(AU) + Cv} \qquad (1.15)$$

High values of f_r indicate considerable thermal inertia and hence thermal storage capacity; see Table 1.7. An assessment of thermal response will include internal partitioning and whether it consists of stud partitioning or block/brick walls.

A building envelope having a thermal response factor of around 3 indicates a lightweight structure. This would include steel frame with stud partitioning and lightweight infill panels in the envelope. A building constructed from concrete/brick/stone with lightweight infill panels in the envelope would be classified as medium weight and have a thermal response of around 6. Values

Table 1.7 Thermal response factor (f_r)

Thermal response factor (f_r)	Recommended preheat times (h)	
	Optimum start	Fixed start
<2.5	2	3
2.5–6.0	3	4
6.0–10.0	4	5
>10	5	6

of f_r between 7 and 10 indicate a structure built entirely with dense materials and having relatively small glazed areas and indicate a building as heavyweight.

As shown in example 1.3, the location of the thermal insulation in the structure is decisive in putting a final value on the thermal response factor (f_r), and can have the effect of modifying the thermal response of an otherwise heavyweight building. It is therefore important to identify the location of the insulation slab in the structure, as it will also affect the admittance Y and hence plant ratio F_3. Look again at equations (1.13), (1.14) and (1.15).

Example 1.6

A heated building has a thermal response factor of 2.5 and is occupied continuously for 8 h/day. If the preheat period before occupancy is 3 h, determine the plant ratio F_3 that must be applied to the design heat loss Q_t for the building of 100 kW and hence find the plant energy output Q_{pb}.

Solution

From equation (1.15) where $H = (8 + 3) = 11$ h:

$$F_3 = \frac{24 \times 2.5}{(11 \times 2.5) + (24 - 11)}$$

from which

$$F_3 = 1.48$$

From equation (1.14):

$$Q_{pb} = 1.48 \times 100 = 148 \, \text{kW}$$

This represents the size needed for the boiler plant. Clearly, if the preheat period is reduced, the plant ratio will go up in value. It should also be remembered that this is the plant load required when outdoor temperature is at design.

The plant ratio (traditionally known as the plant margin or overload capacity) nevertheless does seem to be high but it should be borne in mind that this methodology has been developed with current insulation standards in mind. The effect of these standards has been to reduce significantly the design heat loss and space heater sizes. As a result, following plant shutdown, the smaller space heaters require a large plant ratio F_3 for the preheat period, otherwise the building envelope will take too long to reach optimum temperature prior to occupation, particularly after a weekend shutdown period.

The one exception is where thermal insulation is applied to the room-side surface of the exposed envelope, such as proprietary faced thermal insulation boarding, and here more traditional plant ratios of around 1.1 to 1.25 can be adopted with confidence.

Example 1.7

(a)   A building measures $10 \times 7 \times 3.2$ m high and has four windows each 2×1.5 m. Determine the response factor for the building and obtain the recommended preheat time for a fixed start.

Data

Coefficient	*Wall*	*Roof*	*Floor*	*Glass*	*N air change*
U	0.73	0.4	0.3	5.6	0.75
Y	3.6	2.8	2.5	5.6	

(b)   Given the daily occupancy is 10 h, determine the plant ratio F_3 that will be used for selecting the boiler plant size.

Solution

(a)   See Table 1.8.

$$Cv = \frac{NV}{3} = \frac{0.75(7 \times 10 \times 3.2)}{3} = 55.44 \, \text{W/K}$$

From equation (1.15):

$$f_r = \frac{787 + 55.44}{187 + 55.44} = 3.47$$

From Table 1.3 preheat will be about 4 h and $H = (10 + 4) = 14$ h.

(b)   From equation (1.14):

$$F_3 = \frac{24 \times 3.47}{(14 \times 3.47) + (24 - 14)} = 1.42$$

Table 1.8   Example 1.7: determination of response factor

Element	Dimensions	A	AU	AY
Glass	$4 \times 2 \times 1.5$	12	67.2	67.2
Wall	$(34 \times 3.2) - 12$	96.8	70.7	348.5
Roof	10×7	70	28	196
Floor	10×7	70	21	175
			$\sum (AU)$ 187	$\sum (AY)$ 787

Example 1.8

An office set in its own grounds measures $50 \times 30 \times 3$ m high. Comfort temperature is $20\,°C$ and design outdoor temperature is $-4\,°C$. The air change rate is 0.75 h. There are 20 windows each 2×1.5 m high and two entrance doors each 2×2 m. Occupancy is five days per week from 09.00 to 18.00 hours. The space heating is by panel radiators with optimum start control serving a weather compensated system. Other data are as follows:

Values	Glazing	Wall	Doors	Roof	Floor
U	3.2	0.5	1.2	0.25	$0.3\,\text{W/m}^2\,\text{K}$
Y	3.2	3.7	2.1	0.50	$1.7\,\text{W/m}^2\,\text{K}$

(i)   Determine the building heat loss.
(ii)   Calculate the thermal response factor for the building.
(iii)   Calculate the plant ratio.
(iv)   Determine the boosted plant energy output.

Solution

Element	Dimensions	Area	U	Y	AU	AY
Glazing	$20 \times 2 \times 1.5$	60	3.2	3.2	192	192
Doors	$2 \times 2 \times 2$	8	1.2	2.1	9.6	16.8
Wall	$(160 \times 3) - 68$	412	0.5	3.7	206	1524.4
Roof	50×30	1500	0.25	0.5	375	750
Floor	50×30	1500	0.3	1.7	450	2550
		Total 3480			Total 1232.6	5033.2

Ventilation conductance $Cv = (0.75 \times 50 \times 30 \times 3)/3 = 1125\,\text{W/K}$

From equation (1.10) and taking R for radiators as 0.7 from Table 1.3:

$$F_1 = \frac{3(1125 + (6 \times 3480))}{1232.6 + (18 \times 3480) + 1.5 \times 0.7((3 \times 1125) - 1232.6)} = 1.0$$

From equation (1.11):

$$F_2 = \frac{1232.6 + (18 \times 3480)}{(1232.6 + (18 \times 3480)) + 1.5 \times 0.7((3 \times 1125) - 1232.6)} = 0.966$$

(i)   Building heat loss

$$Q_t = ((1.0 \times 1232.6) + (0.966 \times 1125))(20 + 4) = 55\,664\,\text{W}$$

(ii)   From equation (1.15) the thermal response factor

$$f_r = \frac{5033.2 + 1125}{1232.6 + 1125} = 2.61$$

From Table 1.7, preheat time for optimum start is 3 h and therefore $H = (9 + 3) = 12 \, \text{h}$.

(iii)   From equation (1.14) the plant ratio

$$F_3 = \frac{24 \times 2.61}{(12 \times 2.61) + (24 - 12)} = 1.446$$

(iv)   From equation (1.13) boosted plant output

$$Q_{pb} = 55.7 \times 1.446 = 80.5 \, \text{kW}$$

Conclusions

The boiler plant should be sized at around 81 kW to ensure that after a weekend shutdown during design weather conditions the building envelope is raised to the optimum temperature *and* comfort temperature of 20 °C by 09.00 hours. If the optimum start control is self-learning, it will fine tune the plant start time from its historical data. With the boiler plant sized to deliver 81 kW for design weather conditions at the commencement of the preheat period, it is clear that the system of space heaters must also be able to deliver 81 kW. Example 1.9 investigates this element of design.

Example 1.9

An office has a design heat loss of 56 kW and the selected radiators have a quoted output of 664 W/m^2 at 60 K. Comfort temperature is 20 °C and the flow and return temperatures from the LTHW heating system are 82/70 °C. The radiator manufacturer's index $n = 1.24$ [Caradon Stelrad Elite].

  (i)   Determine the radiator surface required for the offices using a margin of 10%.
 (ii)   Calculate the output from the radiators if the average temperature in the offices were to fall to 10 °C by the early hours of a Monday morning.
(iii)   If the plant ratio is known to be 1.446 and the building thermal response factor is 2.61 find the system flow temperature for boost conditions during the preheat period.

Solution

For radiator output $Q \propto dt^n$ so

$$\frac{Q_2}{Q_1} = \left(\frac{dt_2}{dt_1}\right)^n$$

from which

$$Q_2 = \left(\frac{dt_2}{dt_1}\right)^n \times Q_1$$

Temperature difference dt_2 = mean water temperature − room temperature

Mean water temperature = $0.5(82 + 70) = 76\,°C$

Thus the corrected radiator output for the offices, $Q_2 = ((76 - 20)^{1.24}/60^{1.24}) \times 664$

$$Q_2 = \left(\frac{56}{60}\right)^{1.24} \times 664 = 610\,\text{W/m}^2 \text{ at } 56\,\text{K}$$

Thus as the mean water to room temperature difference decreases, so does the radiator output; and the quoted value of $664\,\text{W/m}^2$ at $60\,\text{K}$ is decreased to $610\,\text{W/m}^2$ at $56\,\text{K}$ for the offices.

$$\text{Net radiator area for the offices } A = \frac{56\,000}{610} = 91.8\,\text{m}^2$$

(i)   Allowing the margin of 10% radiator area, $A = 91.8 \times 1.1 = 101\,\text{m}^2$. When room temperature falls to $10\,°C$, the mean water to room temperature difference

$$dt_2 = 76 - 10 = 66\,\text{K} \quad \text{and} \quad Q_2 = \left(\frac{66}{60}\right)^{1.24} \times 664 = 747\,\text{W/m}^2 \text{ at } 66\,\text{K}$$

(ii)   When the indoor temperature is $10\,°C$, the total output from the radiators will be

$$747 \times 101 = 75\,447\,\text{W}$$

The reader at this point should note that this example is an extension of example 1.8 in which the plant ratio F_3 was calculated to be 1.446; so the plant selected must have an output of $56 \times 1.446 = 81\,\text{kW}$. Clearly the *system* must deliver the same output during the preheat period of $3\,\text{h}$. (Refer to example 1.8.)

The output of the radiators during the preheat period is calculated to be $75.5\,\text{kW}$. The radiator output required to achieve $81\,\text{kW}$ using a surface area of $101\,\text{m}^2$ will be $81\,000/101 = 802\,\text{W/m}^2$. Adopting again the equation $Q_2 = (dt_2/dt_1)^n \times Q_1$ we have

$$802 = \frac{(t_m - 10)^{1.24}}{66^{1.24}} \times 747$$

where t_m will be the required system mean water temperature during preheat. So

$$\sqrt[1.24]{\frac{802 \times 180.4}{747}} = t_m - 10$$

from which $t_m = 80\,°C$.

(iii)   With system flow and return temperatures at 82/70 °C, the system boosted flow temperature during preheat will need to be raised to $80 + (12/2) = 86\,°C$.

Conclusions

With the room temperature at 10 °C the output from the radiators increased from 56 kW to 75.5 kW. The output required from the *system* at this point in time is 81 kW. If the margin on the radiator total area is increased from 10% to 18%, the system output of 81 kW would be met without the need to boost boiler flow temperature. You should now confirm that if this were the case, the revised radiator margin should be 18% by calculating the revised radiator total area from $Q_{pb}/747\,W/m^2$. Alternatively with the radiator area margin at 10% the boiler flow temperature must be boosted from 82 °C to 86 °C during the preheat period as shown in part (iii).

If no margin is applied to the radiator total area at all the required output from the radiators at the beginning of the preheat period would need to be $Q_{pb}/A = 81\,000/91.8 = 882.35\,W/m^2$. Adopting the equation $Q_2/Q_1 = (dt_2/dt_1)^{1.24}$ and substituting $882.35/664 = ((t_m - 10)/(60))^{1.24}$ from which mean temperature $t_m = 85.5\,°C$ and hence boosted boiler flow temperature, $t_f = 85.5 + (12/2) = 91.5\,°C$. This is 9.5 K above the normal boiler flow temperature of 82 °C. You should now confirm that the boosted boiler flow temperature required with no margin on radiator area is 91.5 °C.

The lowest indoor temperature was taken as 10 °C and the solutions were based on this temperature. The indoor frost thermostat setting to start the boiler plant can therefore now be decided.

The above example and its various solutions give an insight to the determination of plant output, plant operating temperatures, and space heating appliance size and variations in output under occupied operating periods and unoccupied periods of preheating. It is important to understand the behaviour of plant and system so that the sizing procedures are right.

Summary to examples 1.8 and 1.9 in which $Q_t = 56\,kW$ and $Q_{pb} = 81\,kW$

Design parameter	Boiler flow & return during preheat (°C)	t_m (°C)	Total radiator area A (m²)	Indoor temp prevailing (°C)	Design outdoor temp (°C)	Preheat time (h)
10% rad. margin	86/74	80	101	10	−4	3
No rad. margin	91.5/79.5	85.5	91.8	10	−4	3
18% rad. margin	82/70	76	108.4	10	−4	3

HIGHLY INTERMITTENT HEATING

Buildings that are occupied for less than 6 h a day and those occupied for less than five days per week will require a local approach to time scheduling for optimum comfort when occupation does occur. Annual energy costs are likely to be out of proportion to the length of the annual occupation period since the building fabric will need reheating before each short period of occupation. The best location for thermal insulation in the building envelope will be on the inside surface as example 1.3 demonstrates. Alternatively high-temperature spot heating in the form of luminous quartz heaters may provide some level of comfort and offset the high radiant heat loss from the human body.

A FINAL COMMENT ON ADMITTANCE

Admittance is determined primarily by the characteristics of the materials in the layers adjacent to the internal surface up to 100 mm. Moving the insulation slab from within the structure to its inner surface will have the effect of reducing the admittance and hence the thermal response factor f_r. The structure's thermal transmittance, however, will remain unchanged unless slabs of material are added or removed.

Conclusions

There are three modes of operation for plant:

- continuous;
- intermittent;
- highly intermittent.

Plant operating modes or time scheduling will depend upon the occupancy levels in the building and upon the outdoor climate.

Clearly, in severe weather when outdoor temperature is below design for extended periods, it may be prudent to ensure that the client is advised to operate continuously or for periods longer than normal, plants that are operated intermittently during normal conditions. The plant ratio F_3 allows the plant to be operated under boost conditions during preheat. That is to say, the system flow temperature may be elevated to increase further the output of the space heating appliances during the preheat period.

If boost conditions during preheat are not a control option, the plant ratio will anyway allow a more rapid warm-up period due to the increased output of the space heaters. This however may not be enough (see example 1.9). The important lesson here is that elevating the boiler thermostat will not achieve greater output if the boiler has been sized on the building heat loss Q_t alone.

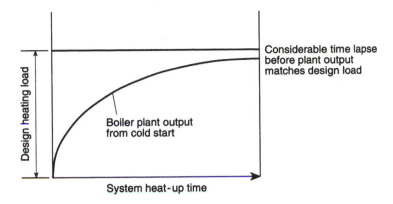

Figure 1.6   The effects of undersized plant on design load.

If the plant ratio/plant margin/overload capacity is not part of the boiler rating, elevating the boiler operating thermostat will result in the plant running continuously. It will not reach its thermostat setting and certainly will not achieve a higher output. Figure 1.6 illustrates the point.

1.4   Sustainable development in building services

All sectors of the building industry now have responsibility in accounting for energy use, useful life, recycling and waste disposal, and the discharge of carbon dioxide into the atmosphere. Energy is also used and carbon dioxide discharged into the atmosphere in the extraction of raw materials, manufacture of products and extraction of fossil fuels. It is used for transport at both ends of the extraction and manufacturing process. Water, another resource, may be used in extraction, manufacture, installation, operating life, recycling and waste disposal.

The building services engineer has a responsibility in the design of systems and selection of plant to ensure that the products used and the systems installed satisfy this accounting process. For example, how much raw material and energy is used to manufacture a heat source like a boiler; how efficient will be its energy use in its lifetime; what will be its working life; how much of it can be recycled; what is the energy and environmental cost of waste disposal; what is the total carbon dioxide emission accruing from its sourcing, manufacture, life cycle, recycle and waste disposal. These are issues that are now challenging product manufacturers, system designers and system users. The footprint of a building with all its services will need to satisfy the checks and balances of sustainability.

Discharge of carbon dioxide from the combustion of fossil fuel in boiler plant is discussed in Chapter 11.

1.5 Chapter closure

This completes the work on heat requirements of heated buildings in temperate climates. You will now be competent to undertake simple thermal modelling of the building envelope, space heater and boiler plant selection, time scheduling and operating parameters which includes determination of building thermal response, preheat times and plant ratio.

Low-temperature hot water heating systems 2

Nomenclature

C	specific heat capacity (kJ/kg K)
CTVV	constant temperature variable volume
CVVT	constant volume variable temperature
d	pipe diameter (m)
dh	difference in head (m)
dp	pressure difference (kPa)
dp_s	identified pressure difference around system (kPa)
dp_v	pressure difference across valve (kPa)
dt	temperature difference (K)
EL	equivalent length for fittings (m)
f	frictional coefficient
F&E	feed and expansion
g	gravitational acceleration (9.81 m/s^2 at sea level)
HTHW	high-temperature hot water
k	velocity pressure loss factor
k_t	total velocity pressure loss factor
K_v	flow coefficient
L	straight pipe length (m)
l_e	equivalent length when $k = 1.0$ (m)
LTHW	low-temperature hot water
M	mass flow rate (kg/s)
MTHW	medium-temperature hot water
MWT	mean water temperature
N	valve authority
pd	pressure drop (Pa, kPa)
Q	output, load (kW)
TEL	total equivalent length (m)
u	mean velocity (m/s)
VFR	volume flow rate (litres/s, m^3/s)
VLTHW	very low-temperature hot water
VTVV	variable temperature variable volume
ρ	density (kg/m^3)
$\sum$	sum of

2.1 Introduction

Water has been used as a liquid heat-carrying medium for a considerable time, owing to its availability and high specific heat capacity of approximately 4.2 kJ/kg K, compared with 2.2 for low-viscosity oil and 1.01 for air.

Organic and mineral liquids are available as liquid heat-transporting media having the advantage of higher evaporation temperatures than water at similar pressures, and minimal corrosion properties, although they have lower specific heat capacities. However, they are more expensive and are therefore rarely used in building services. Table 2.1 identifies arbitrarily the distinctions between water-heating systems operating at different temperatures.

The decision over which level of temperature to use will depend on a number of factors. However, it is important to note that the output of the heat exchanger is dependent upon the magnitude of the difference between mean surface (system) temperature and room temperature, so the higher the mean system temperature is, the greater will be the output of the heat exchanger.

Conversely, the lower the mean system temperature is, the larger will be the heat exchanger surface for the same output. Low surface temperature radiators, for example, are therefore comparatively large.

2.2 Space heating appliances and pipework

It is assumed that you are aware of the common types of heat exchangers on the market and have a knowledge of their construction and operation. If this is not the case, consult manufacturers' literature [*CIBSE Building Services OPUS Design File*].

Heating appliances will employ one or two modes of heat transfer in varying proportions, which you need to identify as this can have a significant effect in the design process. Heat transfer is covered in detail in *Heat and Mass Transfer in Building Services Design* [Spon Press]. Table 2.2 lists some of the more common space-heating appliances. It is important to know where to locate the different types of appliances and what applications are appropriate. Table 2.3 lists some examples.

Table 2.1 Water heating systems

System	Flow temperature (°C)	dt (K)	Supporting pressure (bar gauge)
VLTHW	50	10	–
LTHW	85/80	12/10	–
MTHW	120	15/20	2.7
HTHW	180	30/50	10+

Table 2.2   Common space heating appliances

Indirect	Direct
Radiators	Oil- or gas-fired forced/draught heaters
Natural draught convectors	Radiant vacuum gas heaters
Unit heaters	Oil- or gas-fired radiant heaters
Forced draught convectors	Electric natural draught convectors
Pipe coils	Electric tubular heaters
Embedded pipe coils	Electric underfloor resistance cables
Radiant panels	Electric quartz heaters
Radiant strip	Electric ceramic heaters
Convector radiators	Electric forced draught heaters
Continuous convector	

Table 2.3   Typical applications

Radiators/natural draught convectors:	*Unit heaters:*
Offices	Factories
Schools	Kitchens
Private dwellings	Canteens
Hostels	Garage workshops
Hotels	
Restaurants	
Hospitals	
Forced draught convectors:	*Embedded pipe coils:*
Entrances/foyers	Entrance halls
Assembly halls	Residential
Dining rooms	Public areas
Rooms with limited wall space	Libraries
Stairwells	Museums
Committee/boardrooms	
Radiant strip/panels:	*Luminous heaters (e.g. quartz):*
Factories	spot heating (e.g. churches,
Aircraft hangars	workshops)
Loading bays	

DISTRIBUTION PIPEWORK

The heating medium requires transporting from the heat source to the space-heating appliances. The distribution pipework normally consists of two pipes: a flow and a return. However, there are two other forms of distribution, and a special application of two-pipe distribution (Figure 2.1).

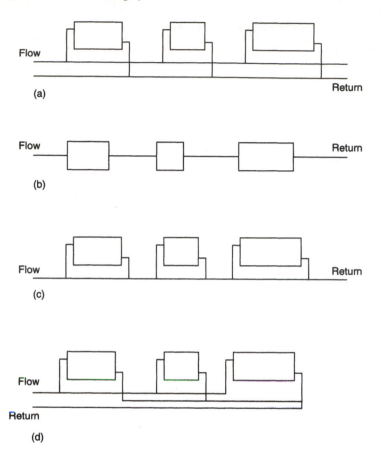

Figure 2.1   Heating distribution pipework: (a) two-pipe distribution; (b) series distribution; (c) one-pipe distribution; (d) reverse return or equal travel.

Two-pipe and reverse return distributions provide a constant mean water temperature throughout the system. Series and one-pipe distributions do not, and system mean water temperature reduces as one moves away from the heat source. This means that appliances must increase in size to maintain the same output. It follows that two-pipe distribution and reverse return, besides being preferable in most cases, offer easier design procedure.

Reverse return uses the most pipework, as three pipes are associated with each appliance. However, it reduces the problems of hydraulic balancing, as each terminal is the same pipe distance from the pump. It is adopted when large numbers of inaccessible heat exchangers are connected, as in the case of a system of air conditioning employing ceiling-mounted fan coil units.

THE HEAT SOURCE

This of course refers to the generator or boiler, of which there are numerous types, and again you are expected to know some of the more common ones or to refer to manufacturers' literature. With fuels such as gas, coal and oil, the boiler provides the facility for combustion to take place with primary and secondary air, and heat transfer by conduction, radiation and convection from the furnace, and flue gases to the water via heat exchangers, which are specially designed and located in the boiler to promote the maximum thermal efficiency from maximum load to a turndown ratio of around 30% of boiler output.

Boiler plants are discussed in further detail in Chapter 6.

THE PRIME MOVER

Circulating pumps are used to transport the water around the system; they are either fixed speed or variable speed (see Chapter 3).

2.3   Pipe sizing

Pipe-sizing procedures for hot pipes are complicated by the effect of pipe emission. This falls into two categories: useful, as in the case of low-level pipes under radiators; and unusable, as instanced by insulated pipework distributed in enclosed ducts or false ceilings.

The effect of pipe emission in a pipe network is to progressively reduce the temperature drop between flow and return as one moves away from the heat source. Thus it might be 12 K at the generator and 7 K at the index terminal. As mass flow rate is determined from

$$M = \frac{Q}{C \, \mathrm{d}t} \quad (\mathrm{kg/s})$$

this also effectively increases the rate of flow, M, progressively from what it would be if $\mathrm{d}t$ was constant at, say, 12 K.

Initially pipe emission is an unknown, as the pipe sizes are unknown. It therefore follows that an estimate of the sizes of pipes in the network must initially be made so as to assess the potential emission. A knowledge of which pipes are lagged and which are unlagged is also required.

Alternatively, pipe emission may be estimated as a percentage of terminal outputs and pipe sizing undertaken as a preliminary process. System hydraulic balancing would then be used to iron out the inaccuracies.

THE INDEX RUN

Although terminals may be connected in series, one-pipe or two-pipe configurations, the space-heating network will consist of circuits in parallel. The
index circuit therefore is that circuit that initially has the greatest pressure
drop: usually the longest circuit in the network.

ALLOWANCE FOR FITTINGS

The hydraulic losses sustained when water is forced around the network result
from that in straight pipes, pipe fittings, control valves, plant and the terminals.
Losses are usually converted into equivalent lengths of straight pipe, at least for
the fittings. This is achieved either by approximation using a percentage on straight pipe or by converting the fittings to equivalent lengths of straight pipe from

$$EL = k_t \times l_e \quad (m)$$

for each pipe section and total equivalent length from

$$TEL = L + EL \quad (m)$$

Values for l_e are obtained in the CIBSE pipe-sizing tables alongside those of
mass flow rate M. Velocity pressure loss factors k for pipe fittings are also
found in the CIBSE pipe-sizing tables.

PRESSURE DROP DUE TO HYDRAULIC RESISTANCE OF STRAIGHT PIPE AND FITTINGS

Pressure drop for each pipe section:

$$dp = (pd/m) \times TEL \quad (Pa)$$

INDEX PRESSURE DROP

This is calculated from the sum of the pressure drops in each pipe section
forming the index run and the hydraulic losses through control valves, plant
and the index terminal on that circuit.

PUMP DUTY

This includes the calculated index pressure drop and the total system flow
rate for which the pump is responsible. Further discussions on pumps are
presented in Chapter 3.

ECONOMIC PRESSURE LOSS FOR PIPE SIZING

For a given rate of flow, the choice of small pipe sizes results in high rates of pressure loss and hence pumping costs. The choice of large pipe sizes for the same rates of flow increases capital costs.

An 'economic' pipe size is said to be achieved at around 300 Pa/m. However, the index run is sized below this at around 250 Pa/m to avoid excessive balancing resistances on the branches. Branches may be sized up to the limiting velocity, which will give pressure drops in excess of 300 Pa/m.

MAXIMUM WATER VELOCITIES IN PIPES

These are up to 1.5 m/s for steel and 1.0 m/s for copper, and above 50 mm diameter, they are 3.0 m/s for steel and 1.5 m/s for copper.

The limiting velocities are in response to the unwanted generation and transmission of noise that can occur at higher values. However, noise generation may not be critical in applications with relatively high background noise levels. Low water velocities give rise to poor signals at commissioning valves.

PIPE-SIZING PROCEDURE

Having calculated the mass flow rates in each pipe section of the network, it is useful to identify each section with a number for reference purposes and then to identify the number of circuits in the system. The sizing procedure below is then followed.

1. Select the index run and index terminal.
2. Size the index circuit.
3. Assess the index pressure drop.
4. Assess the pressure drop at the branches.
5. Size the subcircuits.
6. Assess the balancing resistances.
7. Determine net pump duty.

APPROXIMATE PIPE SIZING

This is sometimes done as a short-cut procedure for pipe sizing by taking a pd of say 300 Pa/m throughout the pipe network and using a constant dt of say 10 K. Table 2.4 shows the heat-carrying capacities for different pipe sizes based on these criteria. Inaccuracies are accounted for in the process of hydraulic balancing. You should now confirm the results in Table 2.4.

Table 2.4 Approximate pipe sizes for transporting heat loads in kW using water at 75 °C, a temperature drop of 10 K and medium grade steel tube

Pipe diameter, d (mm)	Mass flow rate, M (kg/s)	Pressure drop, pd dp (Pa/m)	Velocity, u (m/s)	Load, Q (kW)
15	0.113	300	0.55	4.75
20	0.248	300	0.7	10.42
25	0.462	300	0.75	19.4
32	0.965	300	1.0	40.53
40	1.44	300	1.0	60.48
50	2.69	300	1.2	113
65	5.34	300	1.5	224

If a more detailed analysis is required, the approximate pipe sizes can be identified from Table 2.4 and an estimate made of the pipe emission in each pipe section of the system by taking recourse to the appropriate pipe emission tables in the *CIBSE Guide Sections C3 & C4, or the Concise Handbook*. The detailed analysis can then proceed by apportioning the pipe emission to each pipe section in the network. Have a look at Case Study 2.4.

2.4   Circuit balancing

If a heating system is left unbalanced the design flow rate will not be achieved in the index run, as most of the water will flow through the shortest circuit. The objective, with the assistance of regulating valves, is to make each circuit in the system equal in pressure drop to that in the index run. The pump is not then able to favour any one circuit. This will ensure that design flow rates are achieved in the index circuit and all the subcircuits in the system. This process is also called hydronic balancing.

Case Study 2.1

Figure 2.2 shows diagrammatically an LTHW heating system serving five terminals. Mass flow rates are given as 0.63 kg/s at each terminal, and the pd across each terminal is 5 kPa. The pd through the boiler plant is 20 kPa, and an allowance for fittings on straight pipe of 25% is to be made. Adopting the pipe-sizing tables from the *CIBSE Guide* [*CIBSE Guide Sections C3 & C4, or the Concise Handbook*] for black medium grade tube, follow the sizing procedure outlined above to size and balance the system.

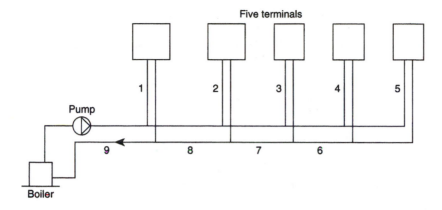

Figure 2.2   Case Study 2.1: LTHW system. Note: cold feed, open vent and F&E tank omitted.

SOLUTION

(1) *Identify the index run:* The lengths of pipe in each section are given in the tabulated data. The index run will consist of sections 5,6,7,8,9. The other circuits consist of the following sections: 4,6,7,8,9; 3,7,8,9; 2,8,9; and 1,9. There are therefore five circuits in the system.

(2) *Size the index circuit:* The index circuit can now be sized at around 250 Pa/m. The results are shown in Table 2.5.

(3) *Assess the index pressure drop:* The index pd can now be determined by summing up the section pressure drops and adding the plant and index terminal pd. The pd in each pipe section is calculated as shown above and included in the table.

$$\text{Sum of section pressure drops} = 18\,878\,\text{Pa}$$
$$\text{Plant and index terminal pd} = 25\,000\,\text{Pa}$$
$$\text{Total index pd} = 43.88\,\text{kPa}$$

Table 2.5   Tabulated results for Case Study 2.1 (medium grade tube)

Section	1	2	3	4	5	6	7	8	9
M (kg/s)	0.63	0.63	0.63	0.63	0.63	1.26	1.89	2.52	3.15
Available pd (Pa/m)	613	448	353	218	250	250	250	250	300
Diameter d (mm)	**25**	**32**	**32**	**32**	**32**	**40**	**50**	**50**	**50**
Actual pd (Pa/m)	515	134	134	134	134	216	152	264	408
Actual length L	8	8	8	8	13	5	5	5	25
TEL (+25%)	10	10	10	10	16.25	6.25	6.25	6.25	31.25
Index pd (kPa)					**2.178**	**1.35**	**0.95**	**1.65**	**12.75**

(4) *Assess the pd at the branches*: The pd available at the branches can now be calculated: considering circuit 4,6,7,8,9, only section 4 requires sizing, and the pressure available at branch 6/4 will be equivalent to that in section 5, namely $2178 + 5000 = 7178$ Pa. As 5 kPa is required at the terminal, the pressure available for sizing section 4 will be

$$pd/m = \frac{pd}{TEL} = \frac{2178}{8 \times 1.25} = 218 \, Pa/m$$

Now taking circuit 3,7,8,9, similarly the pd at branch 7/3 will be equivalent to the sum of those in sections 5 and 6, namely $2178 + 1350 + 5000 = 8528$ Pa. As 5 kPa is required at the terminal, the pressure available for sizing section 3 will be

$$pd/m = \frac{2178 + 1350}{8 \times 1.25} = 353 \, Pa/m$$

For circuit 2,8,9, the pd at branch 8/2 will be equivalent to the sum of those in sections 5, 6 and 7, namely $2178 + 1350 + 950 + 5000 = 9478$ Pa. As 5 kPa is required at the terminal, the pressure available for sizing section 2 will be

$$pd/m = \frac{2178 + 1350 + 950}{8 \times 1.25} = 448 \, Pa/m$$

For circuit 1,9, the pd at branch 9/1 will be equivalent to the sum of those in sections 5, 6, 7 and 8, namely $2178 + 1350 + 950 + 1650 + 5000 = 11\,128$ Pa. Alternatively the pd at this branch will be:

index pd $-$ section 9 pd $-$ plant pd or $43\,878 - 12\,750 - 20\,000 = 11\,128$ Pa

Allowing for 5 kPa at the terminal in section 1 the pd available to size the pipe in this section is given by

$$pd/m = \frac{11\,128 - 5000}{8 \times 1.25} = 613 \, Pa/m$$

(5) *Size the subcircuits*: The available pressure drops can now be entered in the table and the appropriate pipe sections sized accordingly. These pressure drops cannot be exceeded when selecting a pipe size, otherwise one of these subcircuits will exceed the pressure drop in the index run.

This approach to pipe sizing attempts to balance the subcircuits with the index run. If there were an infinite number of pipe sizes from which to choose, the process of hydraulic balancing would be achieved in the choice of pipe size. This has only been possible in section 1, and even here there is a small imbalance. You will notice that the branch sizes are 32 mm except for the branch nearest the pump, which is 25 mm for the same flow rate.

(6) *Assess the balancing resistances*: The balancing resistances can be conveniently obtained as follows:

$$\text{branch 4 pd} = (218 - 134)(8 \times 1.25) = 840 \, \text{Pa}$$
$$\text{branch 3 pd} = (353 - 134)(8 \times 1.25) = 2190 \, \text{Pa}$$
$$\text{branch 2 pd} = (448 - 134)(8 \times 1.25) = 3140 \, \text{Pa}$$
$$\text{branch 1 pd} = (613 - 515)(8 \times 1.25) = 980 \, \text{Pa}$$

These are inserted in the form of regulating valves, which are set accordingly and located in the return pipe of the appropriate branch.

There is a reason for locating commissioning sets in the return rather than in the flow. If the balancing resistance is high, the pressure drop created reduces the antiflash margin on the downstream side of the valve, and if it is located in the flow there is a greater chance of vapourization than there is in the return, which is at a lower temperature.

(7) *Determine the net pump duty*: From the total flow rate and the index resistance the net pump duty will be 3.15 kg/s at 44 kPa. The pump should always be oversized by at least 15% and a regulating valve located on the discharge with an appropriate allowance for its pd in the fully open position. The excess requirement is to allow for variations in pipe routeing to that shown on the drawings, and to have an allowance to assist the commissioning process.

Note: It is a normal practice to install a regulating valve on the final branch to the index terminal and include its pd in the fully open position so that *all* circuits can be regulated. This is necessary for commissioning purposes so that if the circuit chosen at the design stage as the index run is not the *longest* circuit in the system, the commissioning engineer has the facility to select the circuit i.e. much time and money is wasted at the commissioning stage because insufficient attention is given to hydraulic balancing and pump selection at the design stage.

BSRIA have published commissioning codes, which you need to read before attempting the design of a space-heating system.

REDUCING THE NEED FOR BALANCING

If the terminals are connected to a pipe circuit of reverse return, each terminal is the same pipe distance from the pump, and the need for balancing is considerably reduced (see Figure 2.3).

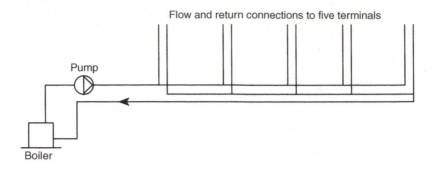

Figure 2.3   Case Study 2.1: system in reverse return.

USE OF BALANCING PIPES

Although we do not have an infinite number of pipe sizes from which to choose during the sizing procedure outlined above, it is possible to balance the branches by selecting a smaller pipe for part of the branch return than that used for the flow, to create a pd equivalent to that required of the regulating valve. It is not normal practice, but it does solve the mystery when you see, on a site visit, differing flow and return pipe sizes serving a terminal or group of terminals.

The generic equation is simple enough:

$$\text{pd in small pipe} + \text{pd in larger pipe} = \text{pd required in branch}$$

$$dp_2 \times L_1 + dp_1(L_2 - L_1) = \text{pd required in branch}$$

where L_1 is the unknown length of balance pipe (m); L_2 is the length of flow plus return in branch (m); dp_1 is the pressure drop for pipe selected for branch (Pa/m); and dp_2 is the pressure drop of pipe next size down (Pa/m).

Consider branch 3 in the case study. The branch size selected was 32 mm at a pd of 134 Pa/m (dp_1). The pd of 25 mm pipe, which is the next size down, is 515 Pa/m (dp_2), and L_2, including the allowance for fittings, is (8×1.25) m.

The pd required in the branch for the purposes of balancing is $353(8 \times 1.25)$. Substituting into the above equation:

$$515L_1 + 134((8 \times 1.25) - L_1) = 353(8 \times 1.25) \qquad .$$

from which

$$L_1 = 5.75 \, \text{m}$$

You should now confirm your agreement. Thus of the total equivalent length of the flow and return (8×1.25) m for branch 3, 5.75 m is required in 25 mm pipe and the rest in 32 mm pipe.

Clearly this can only be a somewhat theoretical solution, as installation rarely follows the working drawing exactly. It may also result in infringing upon maximum water velocities. However, consider the potential savings on a project if, say, a long 100 mm diameter branch main can be partially reduced to 65 mm!

In the above case study mass flow rates were given, thus avoiding the problem of pipe emission. The following case study addresses the issue.

2.5   Apportioning pipe emission

Case Study 2.2

Consider the diagrammatic layout of an LTHW heating system shown in Figure 2.4. Design data are given in Table 2.6. Allowance for fittings on straight pipe shall be 20%. The solution requires the following.

1.   Determine accurately the mass flow rates in each pipe section.
2.   Adopt a suitable index pd and size the index circuit.
3.   Identify the pressures available at the branches and size the branch pipes.
4.   Hydraulically balance the system.
5.   Specify the net pump duty.

SOLUTION

You will have noticed that pipe emission has already been estimated, and terminal and boiler house pressure drops identified. This has been done so that we can concentrate upon the sizing procedure.

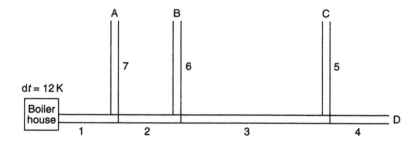

Figure 2.4   Case Study 2.2: schematic layout of lphw heating system in plan.

Table 2.6   Design data for Case Study 2.2

Terminal	A	B	C	D	B/H		
Load (kW)	50	80	30	60	246.7		
pd (kPa)	10	20	10	15	45		
Section	1	2	3	4	5	6	7
Length (flow + return)	30	30	40	20	20	20	20
Pipe emission (kW)	5.6	5.6	6.0	2.5	2.25	2.5	2.25

From Table 2.4, the *CIBSE Guide* [*Section C3 & C4, or the Concise Handbook*] and section pipe lengths given in the case study, two estimates of pipe emission can be made:

- Assuming 25 mm of pipe insulation having a thermal conductivity of 0.07 W/mK, total pipe emission is 5.86 kW, which is 2.7% of the total net load of 220 kW.
- Assuming that the pipe is uninsulated, total pipe emission is 26.7 kW, and this is 12.1% of the total net load.

If the pipe emission is unusable, and it is taken to be so in this case study, it must be added to the net load.

The solution will assume the worst case, namely that the pipe is uninsulated or so badly lagged that its insulating effect must be ignored. The tabulated data reflect these decisions.

(1)   *Accurate determination of the mass flow rates in each pipe section*: The terminal heat loads and estimated pipe emission are annotated on a diagram of the system, as shown in Figure 2.5, for the sake of clarity.

As shown earlier, the formula for mass flow is

$$M = \frac{Q}{C\,dt} \quad (\text{kg/s})$$

M_1 therefore carries the total gross heat load, which is given by

$$\frac{246.7}{4.2 \times 12} = 4.9\,\text{kg/s}$$

At the end of section 1, mass flow is the same but the heat load has changed from 246.7 kW to 241.1 kW. The temperature drop must also

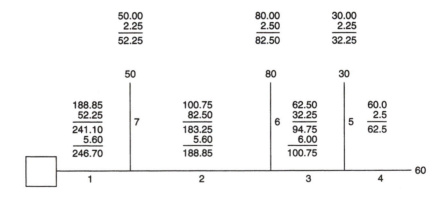

Figure 2.5   Case Study 2.2: diagram of system showing heat load in each pipe.

have changed, and rearranging the formula for mass flow we have at junction 1/2:

$$dt = \frac{241.1}{4.2 \times 4.9} = 11.72 \, \text{K}$$

Thus

$$M_2 = \frac{188.5}{4.2 \times 11.72} = 3.84 \, \text{kg/s}$$

and

$$M_7 = M_1 - M_2 = 4.9 - 3.84 = 1.06 \, \text{kg/s}$$

At the end of section 2 the heat load has changed from 188.85 kW to 183.25 kW, so again the temperature drop has changed, and at junction 2/3:

$$dt = \frac{183.25}{4.2 \times 3.84} = 11.36 \, \text{K}$$

Thus

$$M_3 = \frac{100.75}{4.2 \times 11.36} = 2.11 \, \text{kg/s}$$

and

$$M_6 = M_2 - M_3 = 3.84 - 2.11 = 1.73 \, \text{kg/s}$$

Finally at junction 3/4:

$$dt = \frac{94.75}{4.2 \times 2.11} = 10.7 \, \text{K}$$

and

$$M_4 = \frac{62.5}{4.2 \times 10.7} = 1.39 \, \text{kg/s}$$

with

$$M_5 = M_3 - M_4 = 2.11 - 1.39 = 0.72 \, \text{kg/s}$$

You will have noticed that as we move further away from the heat source the system temperature drop reduces as a result of pipe emission: boiler house $dt = 12 \, \text{K}$, junction $1/2 \, dt = 11.72 \, \text{K}$, junction $2/3 \, dt = 11.36 \, \text{K}$, junction $3/4 \, dt = 10.7 \, \text{K}$. However, if the pipework is efficiently insulated the heat loss is minimal (2.7%), and apportioning pipe emission in this way is necessary only for very long runs of pipe, as in a district heating scheme. As the heat loss is approximately the same in the flow and return, the mean water temperature in a system of two-pipe distribution is considered to be constant.

(2)   *Size the index circuit*: By analysing the data given about the system, one will find that the index circuit consists of pipe sections 1,2,3,4 *or* pipe sections 1,2,3,5. The pipe section lengths in 4 and 5 are similar but the pd in terminal D is greater than that in terminal C: thus the index circuit will consist of pipe sections 1,2,3,4.

With the aid of the *CIBSE* pipe-sizing tables for medium grade tube [*CIBSE Guide Section C4, or the Concise Handbook*] the index pipes can now be sized. This and subsequent information is listed in Table 2.7.

(3)   *Size the branches*: The pd at branch 5 = pd at branch 4 + terminal pd = $6720 + 15\,000 = 21\,720$ Pa.

$$\text{pipe } 5 = 21\,720 - 10\,000 = 11\,720 \text{ Pa}$$

$$dp_5 = \frac{pd_5}{\text{TEL}} = \frac{11\,720}{24} = 488 \text{ Pa/m}$$

The pd at branch 6 = pd at branch 5 + dp_3 = $21\,720 + 9024 = 30\,744$ Pa.

$$\text{pipe } 6 = \text{pd at branch } 6 - \text{terminal pd} = 30\,744 - 20\,000 = 10\,744 \text{ Pa}$$

$$dp_6 = \frac{pd_6}{\text{TEL}} = \frac{10\,744}{24} = 448 \text{ Pa/m}$$

The pd at branch 7 = pd at branch 6 + dp_2 = $30\,744 + 5688 = 36\,432$ Pa.

$$\text{pipe } 7 = \text{pd at branch } 7 - \text{terminal pd} = 36\,432 - 10\,000 = 26\,432 \text{ Pa}$$

$$dp_7 = \frac{pd_7}{\text{TEL}} = \frac{26\,432}{24} = 1101 \text{ Pa/m}$$

Table 2.7   Tabulated results for Case Study 2.2 (medium grade tube)

Section	1	2	3	4	5	6	7
M (kg/s)	4.9	3.84	2.11	1.39	0.72	1.73	1.06
Available pd (Pa/m)	250	250	250	250	488	448	1101
Diameter d (mm)	65	65	50	40	32	40	32
Actual pd (Pa/m)	254	158	188	280	172	425	360
TEL ($L \times 1.2$)	36	36	48	24	24	24	24
pd (Pa) (index)	9144	5688	9024	6720			

These rates of available pressure drop are included in the tabulated results (Table 2.7) and the branch pipes are sized as shown in the table along with the actual rates of pressure drop.

(4) *Determine the balancing resistances*: Using the calculated *available* pds and *actual* rates of pressure drop obtained from the pipe-sizing tables for the pipes selected (see Table 2.7).

$$\text{Branch } 5 = (\text{available pd} - \text{actual pd}) \times \text{TEL}$$
$$= (488 - 172) \times 24 = 7584 \, \text{Pa}$$
$$\text{Branch } 6 = (448 - 425) \times 24 = 552 \, \text{Pa}$$
$$\text{Branch } 7 = (1101 - 360) \times 24 = 17784 \, \text{Pa}$$

(5) *Specify net pump duty*:

$$\text{The index pd} = (1,2,3,4) + \text{BH} + \text{index terminal}$$
$$= 30.576 + 45 + 15$$
$$= 90.576 \, \text{kPa}$$

The total mass flow handled by the pump will be that in section 1, which is 4.9 kg/s. Thus the net pump duty will be 4.9 kg/s at 91 kPa.

NOTES

1. The selected pd for the index circuit is 250 Pa/m. The branch to the index terminal can have a higher rate of pressure loss, as it does not affect the balancing of the branches near the pump.
2. You will notice that the actual rates of pressure drop have been interpolated from the pipe-sizing tables to gain better accuracy.
3. In estimating the pipe emission in the data, approximate pipe sizes would have been obtained from Table 2.4. These now need to be checked with the pipe sizes selected. Any discrepancy requires correction if a full pipe-sizing analysis is needed.
4. The fixed speed pump selected should be at least 15% greater in duty than the net value, to allow for changes during the installation process and a margin for the commissioning engineer within which to work.
5. This case study assumes little or no pipe insulation: hence the need to account for pipe emission, which did not form part of the useful heating surface. If pipework is efficiently lagged, the heat loss from most installations is minimal. This allows a constant temperature drop to be used throughout the system without seriously affecting design accuracy.

    However, where long pipe runs are prevalent, as in the case of district heating, pipe emission must be accounted for.

It can also be accounted for when pipe emission is useful and can be said to form part of the heating surface. If this is the case, the pipe emission does not have to be added to the total plant load.

6. The object in calculating the balancing resistances is to ensure that with them in place in the form of regulating valves set to the appropriate pd, each circuit has the same pressure drop as the index run, namely 91 kPa, and the pump will therefore deliver the design flow to each terminal. This principle is identical to those for balancing ducted air systems.

We have now considered two case studies related to the pipe-sizing process. Case Study 2.3 will look at the pipe-sizing procedure given a fixed pump duty. This situation is not likely to be confronted regularly. It might occur during extensions to an existing installation in which an existing pump is to be commissioned for use.

2.6   Pipe sizing using a fixed pump duty

Case Study 2.3

The pump duty that is available at point X is 4.76 kg/s at 114 kPa. The system to which it is to be connected is shown diagrammatically in Figure 2.6. Further design data are listed in Table 2.8. Allowance for fittings on straight pipe is 15%. Pipework is black heavy grade [*CIBSE Guide Sections C3 & C4, or the Concise Handbook*].

Assume pipework is efficiently lagged and ignore the effects of pipe emission. The design solution requires that the pipework is sized and the system is hydraulically balanced such that the pump having the quoted duty can be installed.

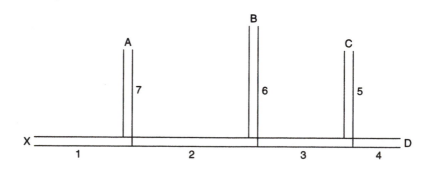

Figure 2.6   Case Study 2.3: diagrammatic layout of system.

Table 2.8   Design data for Case Study 2.3

Terminal	A	B	C	D			
Load (kW)	40	70	60	30			
Terminal pd (kPa)	20	30	25	15			
Pipe section	1	2	3	4	5	6	7
Length (flow + return)	80	100	70	50	60	80	60

SOLUTION

(1)   *Determination of system temperature drop dt*: The total heat load amounts to $\sum(A,B,C,D) = 200\,kW$. Adopting the mass flow rate formula to determine the system temperature drop dt,

$$dt = \frac{Q}{MC} = \frac{200}{4.76 \times 4.2} = 10\,K$$

As pipe emission is to be ignored, this temperature drop will be constant throughout the pipe network.

(2)   *Determination of mass flow rates*: The mass flow can now be determined for each terminal and hence the mass flow calculated for the distribution pipes 3, 2 and 1. These calculations are shown in the results listed in Table 2.9.

(3)   *Terminal pressure drops*: The pds at each terminal can be obtained from the manufacturers of the equipment. These are listed in Table 2.9.

(4)   *Identifying the index circuit*: There are four circuits in the system: 1,7; 1,2,6; 1,2,3,5; and 1,2,3,4. From an analysis of section pipe lengths and terminal pressure drops, circuit 1,2,3,5 is the longest and also can have the greatest pd.

Table 2.9   Tabulated data relating to Case Study 2.3 (heavy grade tube)

Pipe section	1	2	3	4	5	6	7
Load (kW)	200	160	90	30	60	70	40
M (kg/s)	4.76	3.81	2.14	0.714	1.43	1.67	0.95
Terminal pd (kPa)	–	–	–	15	25	30	20
Available pd (Pa/m)	250	250	250	615	250	420	1000
Diameter d (mm)	65	65	50	32	50	50	32
Actual pd (Pa/m)	272	177	228	213	106	140	366
TEL (m)	92	115	80.5	57.5	69	92	69
Section pd (kPa)	25	20.3	18.35	12.25	7.3	12.88	25.25

(5) *Determination of average available pd for the index circuit*: The pump pressure is 114 kPa, so the average rate of pressure drop for the index circuit will be given by

$$\frac{dp}{\text{TEL}} = \frac{\text{available pump pressure} - \text{pd in index terminal}}{310 \times 1.15}$$

$$\text{pd} = \frac{114\,000 - 25\,000}{310 \times 1.15} = 250\,\text{Pa/m}$$

This is added to the tabulated data.

(6) *Sizing the index circuit*: As the pump pressure cannot be exceeded, it is necessary to pay particular attention to the selection of the pipe sizes [*CIBSE Guide Sections C3 & C4, or the Concise Handbook*]. You will see from the tabulated data that one section does exceed 250 Pa/m but the others do not. As a check, the total pd in the index pipe sections can be calculated for the pipe sizes chosen. Using the data in the table, this comes to 71 kPa.

The maximum pump pressure available is $114 - 25 = 89\,\text{kPa}$. If section 5 is reduced to 40 mm pipe, the total pd in the index pipe sections comes to 88.6 kPa. Theoretically this would be acceptable; in practice it would not, as it is too close to the maximum and does not allow for variations between design and installation.

(7) *Sizing the branches*: The pressure available at the branches requires calculation in the same way as in the other case studies above. The available pd for the branch pipes is then determined and this is included in the tabulated data, following which the branch pipes are sized and actual rates of pressure drop recorded, as shown by interpolation from the pipe-sizing tables.

(8) *Determination of the balancing resistances*: From the section and terminal pressure drops tabulated, the unregulated pressure drop can be determined for each circuit:

$$\text{circuit } 1,7 = 25 + 25.25 + 20 = 70\,\text{kPa}$$

$$\text{circuit } 1,2,6 = 25 + 20.3 + 12.88 + 30 = 88\,\text{kPa}$$

$$\text{index circuit } 1,2,3,5 = 25 + 20.3 + 18.35 + 7.3 + 25 = 96\,\text{kPa}$$

$$\text{circuit } 1,2,3,4 = 25 + 20.3 + 18.35 + 12.25 + 15 = 91\,\text{kPa}$$

The pump pressure developed at 4.76 kg/s is 114 kPa, so if a regulating valve is fitted on the pump discharge to absorb $114 - 100 = 14\,\text{kPa}$ then the index circuit will require a regulation of $100 - 96 = 4\,\text{kPa}$, and with the index run now at 100 kPa, the regulation required on the other circuits will be

circuit 1,7 = 100 − 70 = 30 kPa in the return of branch 7

circuit 1,2,6 = 100 − 88 = 12 kPa in the return on branch 6

circuit 1,2,3,4 = 100 − 91 = 9 kPa in the return on branch 4

The regulation on the return on the index run will be 4 kPa. The regulation on the pump discharge will be 14 kPa.

Without this regulation the pump is over-sized, with excess pressure and flow rate. With the rationale developed above, the commissioning engineer therefore has room for adjustment at the pump and in each circuit. This offers the best facility, as the installation will not respond exactly as per the design data. After hydraulically balancing the four circuits in the pipe network beginning with the shortest circuit, which here consists of sections 1,7, the final regulation would be done at the valve on the pump discharge.

2.7   Hydraulic resistance in pipe networks

Having dealt with sizing, proportioning pipe emission and hydraulic balancing, the final section of this chapter focuses on hydraulic resistance in pipe fittings. This matter so far has been considered by using a percentage for fittings on straight pipe. Clearly, the approach can only be an approximation, as the percentage will vary according to the density of fittings in a pipe network. The percentage can vary from 10% for long straight pipe runs to 150% in a boiler plant room, where the density of pipe fittings is high.

Another publication [*Heat and Mass Transfer in Building Services Design*, Spon Press] sets out design from first principles, from which you will see that *head loss* dh due to turbulent flow in pipes consists of three kinds, two of which are considered here:

- that in straight pipes, where d$h = (4 f L u^2)/(2gd)$ m of fluid flowing;
- that in fittings, where d$h = k \times u^2/2g$ m of fluid flowing.

Head loss in pipe fittings is calculated from the product of the kinetic energy expression in the Bernoulli theorem [*Heat and Mass Transfer in Building Services Design*, Spon Press] and the velocity head loss factor k for various pipe fittings given in the *CIBSE Guide* pipe-sizing tables [*CIBSE Guide Sections C3 & C4, or the Concise Handbook*]. The values of equivalent length l_e given in the pipe-sizing tables against each mass flow are for one velocity head, i.e. when $k = 1.0$. It is convenient to continue to express head loss and hence pressure loss through fittings in terms of equivalent lengths of straight pipe. Thus total equivalent length TEL is used as it has been already, and pressure loss dp in a pipe section containing fittings is

$$dp = pd \text{ in Pa/m} \times TEL \quad (Pa)$$

where $TEL = L + (k_t \times l_e)$.

DETERMINATION OF l_e

If therefore head loss in fittings is equated to head loss in an equivalent length of straight pipe l_e, then

$$k \left(\frac{u^2}{2g} \right) = \frac{4fLu^2}{2gd}$$

from which

$$k = \frac{4fL}{d}$$

and rearranging,

$$L = k \times \frac{d}{4f}$$

Therefore

$$l_e = k \left(\frac{d}{4f} \right) \quad (m)$$

and if $k = 1.0$ velocity head,

$$l_e = \frac{d}{4f} \quad (m)$$

Alternatively when $k = 1.0$

$$l_e = \frac{dp_{\text{fittings}} \text{ in Pa}}{dp \text{ in Pa/m}} \quad (m)$$

$$= \frac{(u^2/2g) \times \rho \times g}{dp \text{ in Pa/m}} \quad (m)$$

Example 2.1

Validate the value of equivalent length l_e when $k = 1.0$ for water flowing at 3.20 kg/s in 50 mm black medium grade tube at 75 °C. Data taken from the medium grade pipe-sizing tables [*CIBSE Guide Sections C3 & C4, or the Concise Handbook*]: $\rho = 975$ kg/m^3, d$p = 420$ Pa/m, $u = 1.5$ m/s and $l_e = 2.6$.

Solution

Adopting the second formula:

$$l_e = \frac{(1.5)^2/2g \times 975 \times g}{420}$$

from which

$$l_e = 2.6$$

This equivalent length should now be checked in the pipe-sizing tables [*CIBSE Guide Sections C3 & C4, or the Concise Handbook*].

Example 2.2

Check the value of l_e for water flowing at 14 kg/s in 100 mm pipe at 75 °C. Data from the pipe-sizing tables for heavy grade tube [*CIBSE Guide Sections C3 & C4, or the Concise Handbook*]: $\rho = 975$ kg/m^3, d$p = 240$ Pa/m, $u = 1.7$ m/s and $l_e = 5.9$.

Solution

D'Arcy's formula for turbulent flow in straight pipes:

$$\mathrm{d}h = \frac{4fLu^2}{2gd} \quad \text{(metres of water flowing)}$$

$$\mathrm{d}p = \frac{4fLu^2}{2gd} \times \rho \times g \quad \text{(Pa)}$$

$$\mathrm{d}p/L = \frac{4fu^2\rho}{2d} \quad \text{(Pa/m)}$$

Rearranging for frictional coefficient f:

$$f = \frac{2d(\mathrm{d}p/L)}{4u^2\rho}$$

Substituting data,

$$f = \frac{2 \times 0.1 \times 240}{4 \times (1.7)^2 \times 975}$$

from which

$$f = 0.00426$$

Substituting into $l_e = d/4f$,

$$l_e = \frac{0.1}{4 \times 0.00426} = 5.87 \quad (m)$$

This equivalent length should now be checked in the pipe-sizing tables [*CIBSE Guide Sections C3 & C4, or the Concise Handbook*].

Example 2.3
Part of an LTHW space heating circuit consists of a pipe section having one panel radiator, two 15 mm angle valves and two 15 mm copper bends. Determine, from the data, the pressure loss and the equivalent length of straight pipe for the fittings. Data: $u = 1.0\,\text{m/s}$, $\rho = 975\,\text{kg/m}^3$, $f = 0.00625$, k for radiator $= 2.5$, k for angle radiator valves $= 5.0$ each, k for bends $= 1.0$.

Solution
The total velocity head loss factor $k_t = 14.5$, and from above, $dh = k_t u^2/2g$ m of water flowing. As

$$dp = dh \times \rho \times g \quad (\text{Pa})$$

then

$$dp = k_t \left(\frac{u^2}{2g}\right) \times \rho \times g \quad (\text{Pa})$$

Thus

$$dp = k_t(0.5\rho u^2)$$

Pressure loss generated by water flow through the radiator and fittings by substitution:

$$dp = 14.5 \times 0.5 \times 975 \times (1.0)^2 = 7069\,\text{Pa}$$

Equivalent length $l_e = d/4f$ when $k = 1.0$. Substituting,

$$l_e = \frac{0.015}{4 \times 0.00625} = 0.6$$

Equivalent length of straight pipe $= k_t \times l_e = 14.5 \times 0.6 = 8.7\,\text{m}$

This means that the pressure drop of 7069 Pa through the radiator and fittings in the pipe section is equivalent to that through 8.7 m of 15 mm diameter pipe when water flows at a velocity of 1.0 m/s.

Consider an addendum to this question. Look up the data in the pipe-sizing tables for copper, Table X [*CIBSE Guide Sections C3 & C4, or the Concise Handbook*]. On 15 mm pipe at 1.0 m/s, $M = 0.14$ kg/s and $l_e = 0.6$, which agrees with the calculated value above, and the pressure drop per metre (pd/m) $= 810$ Pa/m. Thus if the length of straight pipe in the section was, say, 6 m, the pressure loss through the pipe *and* fittings would be

$$pd = pd/m \times TEL = pd/m \times (L + (k_t \times l_e)) \quad (Pa)$$

Substituting, pd $= 810 \times (6 + 8.7) = 11\,907$ Pa for that section of straight pipe, the radiator and the fittings. This is the *Standard Method* by which hydraulic resistance is calculated for each pipe section in the network. The first part of the solution determined from *first principles* the pressure drop through the radiator and fittings alone.

The pressure drop of 810 Pa/m of straight pipe from the tables can be checked from first principles by adopting the D'Arcy equation,

$$pd/m = \frac{4fu^2\rho}{2d} \quad (Pa/m)$$

and substituting,

$$pd/m = \frac{4 \times 0.00625 \times 1.0 \times 975}{2 \times 0.015} = 812.5\,Pa/m$$

which is very close to the value of 810 Pa/m taken from the pipe-sizing tables.

Example 2.4
Figure 2.7 shows pipe connections to a unit heater. Determine the pressure drop across A and B from the given data from first principles and by the standard method.

Data
Fittings are malleable cast iron and swept, and include two gate valves and four bends. Connections are in 40 mm black medium weight tube [*CIBSE Guide Sections C3 & C4, or the Concise Handbook*] carrying water at 75 °C, where $M = 1.54$ kg/s, $l_e = 1.9$, pd $= 340$ Pa/m, $u = 1.2$ m/s, $L = 5$ m and $\rho = 975$ kg/m^3.

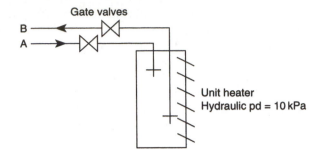

Figure 2.7   Example 2.4: elevation of connections to the unit heater.

Solution from first principles

The numerical data given in the question are obtained from the *CIBSE* pipe-sizing tables [*CIBSE Guide Sections C3 & C4, or the Concise Handbook*] for 40 mm pipe. $l_e = d/4f$, therefore frictional coefficient $f = d/4l_e$. Substituting,

$$f = \frac{0.04}{4 \times 1.9} = 0.00526$$

Hydraulic loss in straight pipe for turbulent flow:

$$dp = \frac{4fLu^2\rho}{2d} \quad \text{(Pa)}$$

Substituting,

$$dp = \frac{4 \times 0.00526 \times 5 \times (1.2)^2 \times 975}{2 \times 0.04} = 1846 \, \text{Pa}$$

The k values for the fittings are

2 gate valves at $0.2 = 0.4$, 4 malleable iron bends at $0.5 = 2.0$

from which

$$k_t = 2.4$$

For fittings:

$$dp = k_t(0.5\,\rho u^2) \quad \text{(Pa)}$$

Substituting,

$$dp = 2.4(0.5 \times 975 \times (1.2)^2) = 1685 \, \text{Pa}$$

The pressure drop between A and B, $dp =$ straight pipe $+$ fittings $+$ terminal

$$dp = 1846 + 1685 + 10\,000 = 13.531\,\text{kPa}$$

Solution by the standard method

The pressure loss between A and B, $dp = (L + k_t \times l_e) \times dp/\text{m} +$ terminal (Pa). Substituting,

$$dp = (5 + (2.4 \times 1.9)) \times 340 + 10\,000 = 13.250\,\text{kPa}$$

The difference between the two solutions is insignificant, but you can see why the standard method is normally used.

The percentage for fittings on straight pipe can be determined for the pipe section, and will be

$$\frac{k_t \times l_e}{L} = \frac{2.4 \times 1.9}{5} = 0.912 = 91.2\%$$

This is high, owing to the density of fittings compared with straight pipe, and is a warning against estimating the percentage unless you have some experience.

2.8   Sizing the temperature control valve

The use of two and three port control valves provides appropriate temperature control to the various circuits in a space heating system. In the selection process, a control valve with a low pressure drop for the required flow rate will give poor control. On the other hand, if a control valve is selected with a high pressure drop for the required rate of flow, control will be good but the pumping costs will be high.

Case Study 2.4

Figure 2.8 represents the plan view of a boiler plant serving an air heater battery with constant temperature variable volume (CTVV) control via a three-port mixing valve providing flow-diverting control. The supply air off the battery is required at constant temperature regardless of the temperature of the air on the upstream side. A thermostat located in the supply air duct therefore controls the three-port valve.

The system (index run) consists of 43.9 m of 32 mm black medium grade tube, 13 malleable iron bends, one straight tee, four gate valves

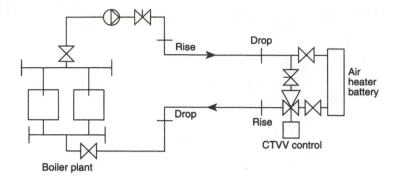

Figure 2.8   Case Study 4: diagram of circuit to a heater battery. Refer to text for pipe lengths and fittings. Cold feed, open vent and F&E tank omitted.

and one regulating valve on the pump discharge. The bypass pipe at the heater battery is 1.5 m long and fitted with a regulating valve. The pipe connections between the bypass and the heater are each 1.2 m in length. The pd across the headers and index boiler is 15 kPa and that across the heater battery is 10 kPa.

From the data, size the three-way valve and determine the net pump duty and the additional pressure drop required across the bypass-regulating valve.

DATA

Tube shall be black medium weight with malleable cast iron swept fittings. Velocity pressure loss factors: gate valve $k = 0.3$, regulating valve $k = 4.0$, bends $k = 0.5$, straight tee $k = 0.2$ and 90° tee $k = 0.5$. Mass-flow rate $M = 1.25$ kg/s, and from the pipe-sizing tables pipe diameter $d = 32$ mm, pd = 500 Pa/m and $l_e = 1.6$ m. Water density at 75 °C is 975 kg/m^3. Three-way mixing valve: valve authority $N = 0.65$. Valve manufacturer's data:

Size (mm)	15	20	25	32
K_v	1.9	4.2	9.6	12.4

INTRODUCTION TO THE SOLUTION

You will note that the valve authority N for the three-way mixing valve has been given in the data. This is required to determine the pressure loss dp_v across the valve when it is fully open to the heater. The selection of control valves involves the use and application of two terms: the *flow coefficient*, sometimes called the capacity index, K_v and the *valve authority N*:

$$\text{VFR} = K_v \sqrt{dp_v} \quad (\text{m}^3/\text{h})$$

from which K_v may be determined.

$$N = \frac{dp_v}{dp_s + dp_v}$$

from which dp_v is calculated.

The flow coefficient K_v is needed to select the control valve from the manufacturer's literature. The valve authority N is needed to determine dp_v. For two-port valves its minimum value is 0.5, and for three-way control valves providing flow mixing, the minimum value for N is 0.3. For three-way valves providing flow diversion, as in this case study, the minimum value for valve authority N is 0.5.

Generally, the higher the valve authority, the better the control but the higher is the pressure drop dp_v across the valve and hence the higher the pump pressure required. Figure 2.9 identifies the terms dp_v and dp_s for three types of valve control.

SOLUTION

The index circuit is that to the heater, which is the index terminal. It does not include the bypass. The total length of straight pipe is 43.9 m.

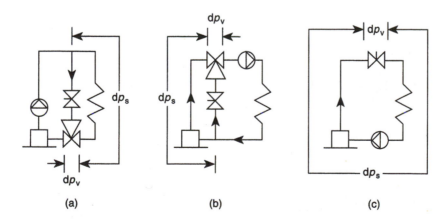

Figure 2.9   (a) Flow-diverting CTVV control; (b) flow-mixing CVVT control; (c) throttling VTVV control.

The total velocity head loss factor k_t for the fittings on the index run includes: 13 bends at $0.5 = 6.5$, 1 straight tee at $0.2 = 0.2$, 4 gate valves at $0.3 = 1.2$, 1 regulating valve at $4.0 = 4.0$ giving $k_t = 11.9$.

$$\text{TEL} = L + k_t \times l_e = 43.9 + 11.9 \times 1.6 = 62.94\,\text{m}$$

Pressure drop in the index circuit

$= \text{TEL} \times dp$ in Pa/m $+$ pressure drop in the index terminal

$+$ pressure drop in the boiler plant

$= (62.94 \times 500) + 10\,000 + 15\,000$

index $dp = 56.47\,\text{kPa}$

However, this *excludes* the pressure drop dp_v across the three-way control valve, which must be accounted for in the net pump duty as it forms part of the index run. From the equation for valve authority,

$$N = \frac{dp_v}{dp_s + dp_v}$$

dp_s represents the pressure loss through the heater circuit downstream of the bypass (see Figure 2.9a) and can be calculated from the data. N is known, so the pressure drop across the control valve can be evaluated and added to the index pd. Rearranging the formula for N in terms of dp_v:

$$\frac{1}{N} = 1 + \frac{dp_s}{dp_v}, \quad \text{then} \quad \frac{1}{N} - 1 = \frac{dp_s}{dp_v}$$

from which

$$dp_v = \frac{dp_s}{(1/N) - 1}$$

$dp_s = $ pd in the pipe connections to the heater downstream of the bypass plus the pd across the heater

$= (L + k_t \times l_e) \times dp\,\text{Pa/m} + 10\,000$

$= (1.2 \times 2 + (2 \times 0.3 \times 1.6)) \times 500 + 10\,000$

$= 11.44\,\text{kPa}$

Substituting N and dp_s into the rearranged formula for N,

$$dp_v = \frac{11.44}{(1/0.65) - 1} = 21.25\,\text{kPa}$$

From the formula for K_v:

$$K_v = \frac{\text{VFR}}{\sqrt{dp_v}} = \frac{(0.00125/975) \times 3600}{\sqrt{0.2125}} = 10$$

From the manufacturer's data, valve size will be 25 mm and $K_v = 9.6$. Substituting this back into the formula for K_v to find the actual pressure drop dp_v,

$$9.6 = \frac{(0.00125/975) \times 3600}{\sqrt{dp_v}}$$

$$\sqrt{dp_v} = \frac{(0.00125/975) \times 3600}{9.6} = 0.481$$

from which $dp_v = 0.23$ bar $= 23$ kPa.

Note: dp_v is in bars and VFR is in m^3/h in the K_v formula.

Thus the *total* index pressure loss will be $56.47 + 23 = 79.47$ kPa. Note that the pressure drop across the control valve is 29% of the total index pd in this system. It cannot therefore be left unaccounted for. Net pump duty will therefore be 1.25 kg/s at 80 kPa. The pump selected must have a duty at least 15% in excess of the net value, for the reasons given earlier.

You will note that the K_v of 9.6 alters the pressure drop through the control valve from 21.25 to 23 kPa. This will affect valve authority N, where $N = dp_v/(dp_s + dp_v) = 23/(11.44 + 23) = 0.67$. The declared value was 0.65.

The second part of the solution relates to hydraulically balancing the bypass with the heater circuit downstream of it. When the heater is on full bypass the pressure drop must be similar to that in the index circuit. If the pd across the control valve when it is on full bypass is the same as when it is fully open to the heater, the hydraulic resistance through the bypass pipe will include: straight pipe, a 90° tee and the fully open regulating valve. Thus

$$pd = (L + k_t \times l_e) \times dp \quad (\text{Pa/m})$$

$$k_t = \text{reg valve} + 90° \text{ tee} + \text{tee bend} = 4.0 + 0.5 + 0.5 = 5.0$$

and

$$pd = (1.5 + (5 \times 1.6)) \times 500 = 4.75 \text{ kPa}$$

The pd through the heater circuit is 11.44 kPa, so the balancing resistance required of the regulating valve is $(11.44 - 4.75) = 6.69$ kPa.

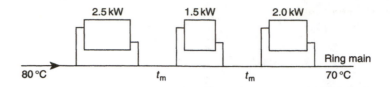

Figure 2.10   Case Study 2.6: one-pipe system.

Both circuits – that through the heater and that through the bypass – now have similar pressure drops each of 79.47 kPa and the system will operate hydraulically as specified.

2.9 Underfloor heating

Underfloor heating systems using VLTHW in copper, steel or plastic piping are now popular. This is largely due to the increase in levels of thermal insulation in the building envelope which means that the system can adequately offset the design heat loss. In older buildings, where underfloor heating was used, additional supplementary heating was required to account for the heat loss.

Case Study 2.5

Figure 2.11 shows a section through an intermediate floor in which an embedded pipe coil is located. The pipe coil is laid at 250 mm centres

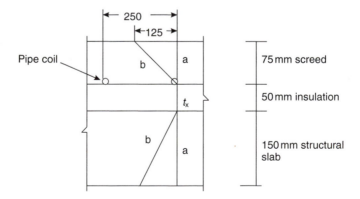

Figure 2.11   Showing section through the floor slab and the location of the pipe coil in the screed. Maximum floor temperature will be vertically above the pipe. Minimum floor temperature will be 125 mm from each pipe. Mean floor temperature of 26 °C will occur at point (a + b)/2.

in 75 mm of screed having a thermal conductivity of 0.4 W/mK. The thermal insulation under the pipe coils is 50 mm thick with a thermal conductivity of 0.05 W/mK and the supporting structural concrete floor under is 150 mm thick with a thermal conductivity of 1.5 W/mK. The combined heat transfer coefficient at the floor surface is 10 W/m²K and at the ceiling surface below 8 W/m²K. Room comfort temperatures above and below are 20 °C. The average floor temperature t_m above the pipe coil shall not exceed 26 °C.

(i) Determine the flow and return temperatures required of the coil if circuit temperature drop is 8 K;
(ii) Estimate the upward heat flux of the pipe coil in W/m²;
(iii) Estimate the downward heat flux of the pipe coil in W/m²;
(iv) List the advantages and disadvantages of this type of space heating.

Note: A detailed account of the combined heat transfer coefficient h is given in another publication [*Heat and Mass Transfer in Building Services Design*, Spon Press]. It is also known as surface conductance h_s, and $h_s = 1/R_s$ where R_s is the surface thermal resistance from which $R_s = 1/h_s$ (m² K/W). Heat flux I is also detailed in the same publication where $I = (t_s - t_c)/R = h(t_s - t_c)$ (W/m²).

SOLUTION

(i) Assuming steady temperatures, a heat balance may be drawn such that:

Mean heat flow from pipe to floor surface
= mean heat flow from floor surface to room.

Thus $(t_m - t_c)/R = h(t_f - t_c)$ where mean thermal resistance $R = L/k$ and mean thickness L from Figure 2.11 will be $(a + b)/2$. This represents the mean distance from the pipe coil to the floor surface and therefore relates to the mean temperature of the floor surface of 26 °C.

$$\text{Dimension } b = \sqrt{(75)^2 + (125)^2} = 146 \text{ mm.}$$

So $R = (a + b)/2k = (0.075 + 0.146)/(2 \times 0.4) = 0.27625$ and $(t_m - 20)/0.27625 = 10(26 - 20)$, from which

mean water temperature, $t_m = 42.6 \,°C$

So

circuit flow temperature, $t_f = 42.6 + 8/2 = 46.6 \,°C$

and

$$\text{circuit return temperature, } t_r = 42.6 - 8/2 = 38.6\,^\circ\text{C}$$

(ii)   Upward emission $= h(t_f - t_c) = 10(26 - 20) = 60\,\text{W/m}^2$. This is about the maximum mean heat output from a floor with embedded pipe coils. Floor temperatures much in excess of $26\,^\circ$C tend to lead to discomfort for sedentary occupation.

(iii)   The first step here is to find the temperature t_x at the interface of the insulation and the structural floor. Using a heat balance,

maximum heat flow from pipe coil to t_x

$=$ maximum heat flow from t_x to the room below

Thus $(t_m - t_x)/R = (t_x - t_c)/R$ and substituting,

$$\frac{42.6 - t_x}{0.05/0.05} = \frac{t_x - 20}{(0.15/1.5) + (1/8)}$$

so $(42.6 - t_x) = (t_x - 20)/0.225$
from which

$$t_x = 24.15\,^\circ\text{C}$$

A further heat balance can now be adopted to find the mean surface temperature t_s of the ceiling immediately below the pipe coil:

Mean heat flow from t_x to $t_s =$ mean heat flow from t_s to t_c

Thus

$$\frac{t_x - t_s}{(a + b)/2k} = h(t_s - 20)$$

where $b = \sqrt{((150)^2 + (125)^2)} = 195\,\text{mm}$. Substituting,

$$\frac{24.15 - t_s}{(0.15 + 0.195)/2 \times 1.5} = 8(t_s - 20)$$

then $(24.15 - t_s) = 0.92(t_s - 20)$. So $24.15 - t_s = 0.92t_s - 18.4$; from which

$$t_s = 22.16\,^\circ\text{C}$$

Downward emission will be $= h(t_s - t_c) = 8(22.16 - 20) = 17.3\,\text{W/m}^2$

Total emission up and down from the pipe coil $= 60 + 17.3 = 77.3\,\text{W/m}^2$

Proportion of heat emission up $= 60/77.3 = 77.6\%$

Proportion of heat emission down $= 17.3/77.3 = 22.4\%$

(iv)   Advantages of underfloor heating: Since the system provides low temperature radiant heating the vertical room temperature profile is constant; this can provide one of the best levels of thermal comfort. Floor space is not taken up with space heaters. A seasonal heating and cooling facility is possible with warm water flow in winter and cold water flow in the summer. No supplementary heating should be necessary. The system is classified as VLTHW and will take full advantage of condensing boiler plant. Consideration could be given to connecting the system to a heat pump due to the low operating temperatures. Pipes can be embedded in a variety of floor constructions in new and existing buildings. You should consult specialist installers' literature for details.

(v)   Disadvantages of underfloor heating: Thermal insulation of the building envelope must be to current standards. Floor finish requires careful selection to ensure against the effect of unwanted thermal insulation. The floor mass above the underlying thermal insulation where it consists of concrete screed in which the pipes are laid may result in a slow thermal response time. See Chapter 1. Careful site supervision is required during installation of the pipe coils. In multistorey buildings heated in this way some heat from the floor coil will dissipate into the rooms below. See solution (iii).

2.10   One pipe radiator systems

Figure 2.1c illustrates one-pipe distribution, which is sometimes used for radiator systems. The ring main connecting the radiators is a common-sized pipe for ease of installation. This has a significant effect upon the ratio of water flow *through* each radiator compared with that through its associated bypass pipe. As the hydraulic pressure drop through the radiator bypass must equal the pressure drop through the radiator it follows that the majority of water in the ring main will flow through each radiator bypass. The ratio M_t/M_r, which is the total mass flow in the ring main to the mass flow through the radiator, is required to determine the temperature drop across the radiator and hence its output in watts.

The method for determining the M_t/M_r ratio is based on trial and error by giving values to ratio M_t/M_r, radiator pipe connection sizes d and lengths, radiator length L_r, ring main size D, and ring main flow rate M_t taken at a pressure drop of between 200 Pa/m and 300 Pa/m. The hydraulic resistance through the radiator and through its bypass can then be checked, and if they are approximately equal the ratio is validated.

Table 2.10 summarizes some validated M_t/M_r ratios for sizing one-pipe systems for radiators having two angle-type valves and 1 m of flow and return pipe.

Table 2.10 M_t/M_r ratio for sizing one-pipe systems

d	D	M_t/M_r ratio where $L_r =$					Approx. value
		1.0m	1.3m	1.6m	2.0m	2.3m	
15	15	3.46	3.19	3.02	2.88	2.76	3
	20	5.92	5.49	5.18	4.95	4.71	5
	25	9.62	9.01	8.47	8.06	7.41	9
20	20	3.72	3.48	3.31	3.17	3.05	3.5
	25	5.75	5.41	5.13	4.90	4.55	5
	32	9.90	9.43	8.93	8.62	8.26	9
25	25	3.95	3.75	3.57	3.44	3.19	3.5
	32	6.54	6.21	5.95	5.71	5.52	6
	40	8.77	8.40	8.06	7.75	7.52	8

Case Study 2.6

Consider the one-pipe system shown in Figure 2.10 and determine:

(a) mean water temperature (MWT) in each radiator;
(b) correction factor for each radiator for sizing purposes;
(c) size of the ring main and radiator connections.

DATA

Radiator outputs given for a temperature difference between radiator and room of 60 K; room temperature 20 °C; ignore pipe emission; black medium-weight pipe shall be used.

SOLUTION

The required radiator outputs are based on the design heat loss calculations. A radiator manufacturer's brochure will specify outputs in watts, usually for a temperature difference between radiator and room of 60 K. For other values of this temperature difference a correction factor must be applied. This is obtained from the following formula:

$$\text{Correction factor } cf = ((\text{design } dt)/(\text{brochure } dt))^n$$

where index n is approximately equal to 1.3 for radiators.

The incremental temperature drop across the ring main $dt_m = dt_t(q/Q)$. The temperature drop across the radiator, dt_r, is obtained from

$$dt_r = dt_t \left(\frac{q}{Q}\right)\left(\frac{M_t}{M_r}\right)$$

where dt_t is the total temperature drop for the circuit, q is the radiator output and Q is the total radiator output for the circuit.

The mass flow in the ring main is

$$M_t = \frac{Q}{C \cdot dt} = \frac{6}{4.2(80 - 70)} = 0.143 \, \text{kg/s}$$

If 20 mm pipe is used, the pressure drop from the pipe sizing tables for medium grade tube [*CIBSE Guide Sections C3 & C4, or the Concise Handbook*] is 107 Pa/m, and from Table 2.10 the approximate M_t/M_r ratio is 5 for 15 mm radiator connections. Thus for the first radiator,

$$dt_r = (80 - 70)\left(\frac{2.5}{6}\right)(5) = 20.8 \, \text{K}$$

This is too high, so the radiator connections need to be increased from 15 mm to 20 mm, for which $M_t/M_r = 3$ and

$$dt_r = (80 - 70)\left(\frac{2.5}{6}\right)(3) = 12.5 \, \text{K}$$

which is satisfactory.

For the first radiator therefore, the mean water temperature = $80 - 12.5/2 = 73.75\,°\text{C}$ and the mean water to room temperature difference $= 73.75 - 20 = 53.75 \, \text{K}$. Correction factor for the first radiator $cf = (53.75/60)^{1.3} = 0.867$.

For the second radiator,

$$dt_m = dt_t\left(\frac{q}{Q}\right) = (80 - 70)\left(\frac{2.5}{6}\right) = 4.17 \, \text{K}$$

and

$$t_m = 80 - 4.17 = 75.83\,°\text{C}$$

This will be the flow temperature to the second radiator. The temperature drop across this radiator for 15 mm connections,

$$dt_r = dt_t\left(\frac{q}{Q}\right)\left(\frac{M_t}{M_r}\right) = 10\left(\frac{1.5}{6}\right)(5) = 12.5 \, \text{K}$$

and the mean water temperature in the second radiator = $75.83 - 12.5/2 = 69.58\,°\text{C}$ and mean water to room temperature difference for this radiator = $69.58 - 20 = 49.58 \, \text{K}$. Correction factor for this radiator $cf = (49.58/60)^{1.3} = 0.78$.

For the last radiator,

$$dt_m = dt_t\left(\frac{q}{Q}\right) = 10\left(\frac{1.5}{6}\right) = 2.5 \, \text{K}$$

Table 2.11 Results for Case Study 2.5

Radiator	Mean water temperature (°C)	Correction factor	Radiator Connections (mm)	Ratio M_t/M_r	Radiator dt (K)
First	73.75	0.867	20	3	12.5
Second	69.58	0.780	15	5	12.5
Third	73.33	0.688	15	5	16.7

Ring main size: 20 mm having a pd of 107 Pa/m.

and

$$t_m = 75.83 - 2.5 = 73.33\,°C$$

This will be the flow temperature to the third radiator.

The radiator temperature drop using 15 mm pipe connections,

$$dt_r = dt_t \left(\frac{q}{Q}\right)\left(\frac{M_t}{M_r}\right) = 10\left(\frac{2}{6}\right)(5) = 16.7\,K$$

and the mean water temperature in this radiator $= 73.33 - 16.7/2 = 64.98\,°C$ and the mean water to room temperature difference for this radiator $= 64.98 - 20 = 44.98\,K$. Correction factor for this radiator $cf = (44.98/60)^{1.3} = 0.688$.

The results from the solution are tabulated in Table 2.11.

CONCLUSIONS

Note the effect on radiator mean water temperature in the system. It reduces in each successive radiator, requiring the calculation of the correction factor each time. The mean water temperature in a system of two-pipe distribution is constant, requiring only one calculation of the correction factor, assuming all rooms are at the same temperature.

Having determined the correction factors, each radiator can now be sized from the manufacturer's brochure.

2.11 Chapter closure

This concludes the work on pipe sizing and hydraulic balancing, proportioning pipe emission, determination of hydraulic resistance in pipe networks from first principles and by adopting the standard method and determination of net pump duty.

You now have the skills to undertake the principle design procedures for sizing wet space heating systems.

Pump and system 3

BPV	balanced pressure valve
CVVT	constant volume variable temperature
D	impeller diameter (m)
F&E	feed and expansion
g	gravitational acceleration (m/s^2)
h	suction lift, head (m)
K	constant
M	mass flow rate (kg/s)
N	speed (rev/s)
NP	neutral point
NPSH	net positive suction head
P	pressure (Pa, kPa)
pd	pressure drop (Pa, kPa)
P_w	power (W)
Q	volume flow rate (litres/s, m^3/s)
TRV	thermostatic radiator valve
+ve, −ve	positive, negative pump pressure
ρ	density (kg/m^3)
η	pump efficiency

3.1 Introduction

For pipe networks distributing water and requiring a prime mover, the centrifugal pump is well established. The pressure that the pump must develop is dependent upon its application.

There are essentially two types of pipe distributions in which a pump is used to transport water. *Closed systems* include those applications where water is recirculated around the pipe network for the purposes of transferring heat. *Open systems* are those in which water is circulated and at some point or at a number of points around the pipe network it is discharged to atmosphere for the ultimate purpose of consumption.

3.2 Closed and open systems

CLOSED SYSTEMS

For closed systems such as space heating, the net developed pump pressure must equal the hydraulic resistance in the index run. It does not matter if the system serves a multi-storey building, as once water movement is initiated, circulation will take place regardless of the static height of the system. However, in such systems, the pump must generate sufficient pressure *at no flow* (i.e. maximum pressure development) to overcome the system's static height.

OPEN SYSTEMS

In the case of open systems such as boosted water from a ground storage tank, the net developed pump pressure must include:

index pd + pressure equivalent to the static lift +
discharge pressure at the index terminal

Open systems: special cases

Where a centralized hot water supply system is connected to a high-level cold water storage tank and relies on the secondary pump to overcome hydraulic resistance, the net pump pressure required includes:

index pd + discharge pressure at the index terminal

The static lift for the system is accounted for by the pressure generated by the static height of the water storage tank.

If booster pumps are connected directly to the incoming water main to provide water to a multi-storey building, pressure required includes:

index pd + pressure equivalent to the static lift − minimum mains
pressure + discharge pressure at the index terminal

3.3 Pump considerations

SUCTION LIFT

Consider the pump connected to an open system in Figure 3.1. It is clear which part of the system is under negative pump pressure (pump suction) and which is under positive pump pressure. The *change* in pump pressure occurs at the impeller within the pump casing. This is the *neutral point* in the open system.

If the pump is drawing water from some point below the centre line of the impeller, the vertical distance h must not be sufficient to cause the water to separate and vapourize in the suction pipe or on the impeller surface.

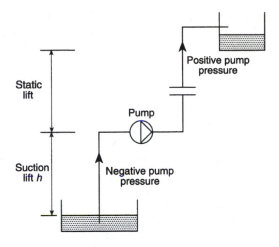

Figure 3.1   Pump pressure effects: open systems.

Theoretical maximum suction life h for cold water is obtained from

$$P = h \times \rho \times g \quad (\text{Pa})$$

from which

$$\text{suction lift } h = P/\rho g \quad (\text{metres of vertical height})$$

If all of atmospheric pressure is used:

$$h = \frac{101\,325}{1000 \times 9.81} = 10.33\,\text{m}$$

This is the maximum theoretical suction lift, and if it were possible to achieve it, the water at the top of the lift would be vapour and at this point we would have also achieved a perfect vacuum. The pump therefore would be attempting to handle vapour, which it is not designed to do. The practical suction lift would be 4 m maximum where the saturation temperature of the water at the top of the lift would be about 76 °C. *Net Positive Suction Head* (*NPSH*) is a term used by pump manufacturers:

NPSH = pressure required at the eye of the impeller to
prevent cavitation in the pump casing

PUMP PRESSURE DISTRIBUTION AND THE NEUTRAL POINT

In open systems where the water level is below the pump, the pump pressure is negative in the suction pipe and positive in the discharge pipe. The pressure change occurs at the impeller. This is called the *neutral point*.

With closed systems and pumped HWS when there is no draw-off, the pump pressure effect changes from negative to positive or vice versa at the neutral point, which is normally taken at the feed and expansion pipe entry into the system.

PUMP LOCATION

This is important for both closed and open systems. In open systems the suction lift must not be excessive, in case cavitation is promoted in the suction pipe or pump casing. With closed systems, pump location influences the water level in the open vent pipe in the case of vented heating systems. It also influences the magnitude of the antiflash margin around the heating system, whether it is pressurized or vented.

ANTIFLASH MARGIN

The antiflash margin is the temperature difference between the water and its saturation temperature at a point in a heating system. The minimum acceptable value is 10 K, with 15 K being preferred.

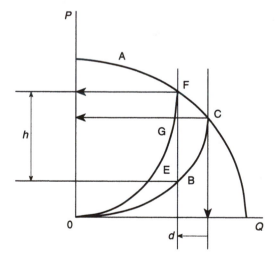

Figure 3.2   The regulating valve produces the required pd (h) to achieve design flow (d). (A) pump characteristic; (B) system characteristic; (C) operating point of the pump on the system; (d) design flow rate; (E) original design conditions for the system; (F) final operating point of the pump on the system; (G) final location of the system characteristic; (h) required pd across the regulating valve to achieve design flow.

PUMP SIZING

At the risk of repetition, as this point has been documented elsewhere, it is essential for the purposes of commissioning to ensure that the pump is over-sized by at least 15%. A regulating valve, which can be located in the pump discharge, is used to move the operating point of the pump on the system up the pump characteristic (from C to F) to the design flow rate (d) (Figure 3.2).

You will notice that the final pump pressure developed is the sum of the original design system pressure (E) plus the pressure drop (*h*) required across the regulating valve.

MULTIPLE PUMPING

Identical pumps connected in series double the pressure developed. Identical pumps connected in parallel double the flow (Figures 3.3 and 3.4). Dissimilar pumps may only be connected in parallel.

CHOICE OF PUMP

For pumps with steep characteristics, changes in pressure developed produce only small changes in flow rate. This is useful where pipes tend to scale up.

For pumps with flat characteristics, changes in flow produce small changes in pump pressure developed. This might be useful where extensive hydraulic balancing is needed, as changes in flow rate will be small in the operating range of the pump.

For closed systems the pressure developed at zero flow – i.e. maximum pump pressure – should be greater than the static height of the system to

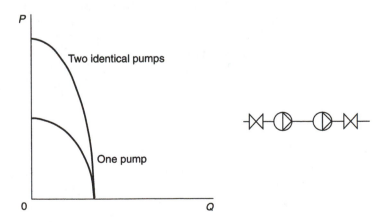

Figure 3.3   Similar pumps connected in series.

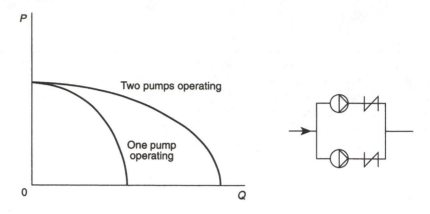

Figure 3.4   Similar pumps connected in parallel.

ensure initiation of flow. Selection should be based upon the intersection of pump and system characteristic at design flow at the point where the pump efficiency curve is at or near its maximum.

PUMP TYPE

There are a variety of pump types: reciprocating, rotary gear, rotary moyno screw, centrifugal single stage, centrifugal multistage, inline and wet rotor. There are also different motor/impeller coupling arrangements: direct coupled, close coupled, belt drive and wet rotor which follows a rotating magnetic field.

Each of these pump types has a traditional application. Reciprocating pumps, for example, were used exclusively for steam boiler feed because they provide positive displacement. Rotary gear pumps are used to pressurize oil feed lines. Wet rotor pumps tend to be used on domestic and small commercial heating systems. Multistage centrifugal pumps are now used exclusively where high pressures are required, as in the case of boosted cold water systems.

The single-stage centrifugal and inline pumps currently have the major market share in the services industry. Recourse should be made to current manufacturers' literature if your knowledge of pumps is limited, before investing further time in this study.

3.4   Pumps on closed systems

This chapter will focus now on circulating pumps for space heating systems, both pressurized and vented. As already stated, the net pressure developed by the pump must just equal the hydraulic resistance in the index run, which is that circuit in the system having the greatest pressure drop. Often it is the longest circuit in the system. Pump capacity is the total flow rate in that part of the system that the pump is serving. Pump duty must specify flow rate *and* pressure developed. (Pumps on open systems are considered in Chapter 9.)

THE NEUTRAL POINT

Consider the system shown in Figure 3.5. When the pump runs, some water is removed from the feed and expansion tank and is deposited in the open vent. This causes a noticeable rise in water level in the open vent, but the corresponding fall in water level in the tank is negligible, owing to the difference in cross-sectional areas. Thus the pressure change at the cold feed entry to the system is also negligible, whether the pump is operating or not. This point is known as the *neutral point* in the system. It is here that pump pressure passes through zero from positive to negative pressure. From this knowledge it is now possible to identify the positive and negative pump effects around the system.

The systems shown in Figures 3.6 and 3.7 identify different arrangements for pump and cold feed location and the positive and negative pump effects at different points around each system.

Clearly, when the pump operates, the system is under the algebraic sum of static pressure and pump pressure at any point. It is this combined pressure that is measured on a pressure gauge. A matter for investigation occurs when at a point in a system static pressure is low (usually high-level pipework) and *negative pump pressure is high*, as combined pressure or gauge pressure at that point may be sub-atmospheric. This will have the effect of reducing the anti-flash margin, and there is then the potential for the water at that point to boil.

Which of the four systems in Figures 3.6 and 3.7 is likely to produce this effect? Which system offers the safest pump location? Note the location of

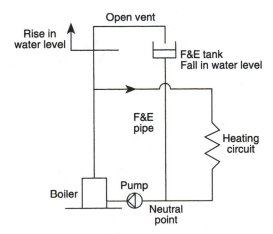

Figure 3.5   Identifying the neutral point in a closed system.

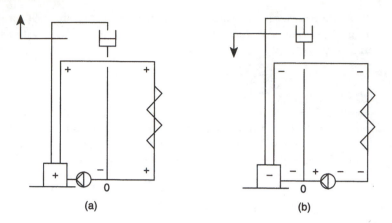

Figure 3.6   Positive and negative pump effects, pump in the return: (a) F&E pipe near pump suction; (b) F&E pipe near pump discharge.

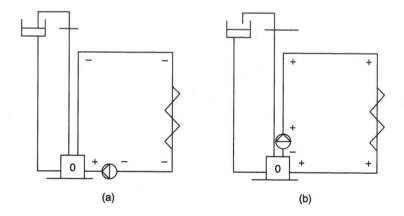

Figure 3.7   Positive and negative pump effects, F&E pipe connected to boiler: (a) pump located in boiler return; (b) pump located in boiler flow.

the open vent connection and cold feed entry. Which is the best engineering solution?

An analysis of the pump effects in each system allows the following conclusions to be drawn:

1.   Negative pump pressure in high-level circulating pipes should be avoided, as it reduces the antiflash margin.
2.   Negative pump pressure in the boiler, where static head is low (rooftop plant rooms are a case in point), should be avoided for the same reasons. One way of overcoming this is to pressurize the system.
3.   Positive pump pressure at the open vent may cause discharge.

4. Severe negative pump pressure at the open vent may cause air to be drawn into the system at the open vent connection to the system.
5. The open vent should not be extended through the roof to offset the effects of point 3 in case of freezing.
6. Positive pump pressure in the system assists venting. If this is the case, the pump can be operated during system venting.
7. Negative pump pressure in the system does not assist venting – it may allow air in – and therefore it should not be operational during this procedure.
8. The open vent is a safety pipe in case the operating boiler and limit thermostats fail. In this event the boiler water will eventually cavitate, as the burner will run continuously. The open vent therefore must be connected directly to the boiler so that steam can escape in these circumstances.
9. The cold feed and expansion pipe is used for initially filling the system and during heat-up allows expansion water to rise into the F&E tank. Thus during normal operation of the system the ball valve in the tank is submerged. This can cause it to stick in the closed position owing to the upthrust on the ball float, and during the summer, make-up water cannot replace water in the system, which evaporates.

As the expansion takes place at the heat source, which is the boiler, the F&E pipe should be connected to or close to the boiler.

You now have the skills to ensure an engineering design solution with respect to optimum pump/cold feed/open vent location. You should also be able to diagnose faults relating to the potential problems described here, in an existing installation, and offer suitable remedies.

USE OF THE REGULATING VALVE

If the pump is oversized as it should be, the operating point of the pump on the system will give a flow rate greater than design flow. The regulating valve, usually located on the pump discharge, is used to throttle the flow down to the design value on completion of the process of hydraulic balancing. The effect is to move the system characteristic up the pump curve from the operating point where the two characteristics intersect to the new intersection and final operating point. This increases the pump pressure developed by adding the pressure drop required at the regulating valve to achieve design flow (Figure 3.2).

SPEED REGULATION

Fixed speed pumps must not operate at no flow conditions in a pipe circuit. If this is likely to occur during system operation a bypass in the circuit is required. The use of two-port on/off control valves or thermostatic radiator

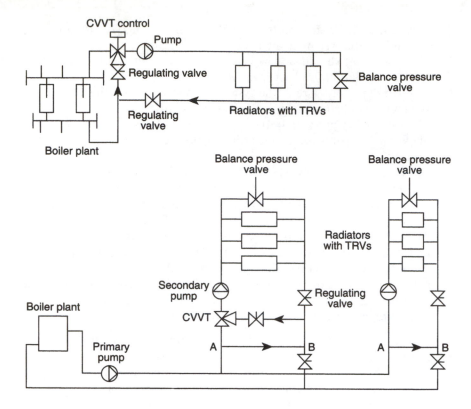

Figure 3.8 Application of balanced pressure valves in circuits subject to interruption of water flow (open vents, cold feeds and F&E tanks omitted).

valves in a circuit are examples. The bypass pipe will be fitted with a balanced pressure valve (BPV) to ensure that water can flow in the pump circuit at all times when the system is in operation (Figure 3.8).

Bypass pipe A–B prevents the high differential pressure in the main distribution being transmitted to the circuit. The BPV, which is spring loaded and normally closed when the TRVs are open, ensures that when all the TRVs are closed, the pump can still pump water through the circuit. However as soon as just one TRV starts to close, the other radiators will begin to receive more flow than they require and the circuit becomes unbalanced. This situation is sometimes ignored in design. When it is accounted for, the use of a differential pressure controller located in the flow and controlling a modulating valve in the return ensures that flow through the circuit and the pressure drop across it is correct at all times.

The alternative use of a variable-speed pump, operating on pressure signals, provides for savings in pumping costs during partial circuit load and overcomes some of the difficulties outlined above as long as a bypass is put across the index terminal in the circuit.

3.5 Centrifugal pump laws

Note: This topic is investigated in detail in another publication. [*Heat and Mass Transfer in Building Services Design*, Spon Press] For centrifugal pumps there is a relationship between Q, N, P, P_w and D such that

$$Q \propto N, \quad P \propto N^2, \quad \text{therefore } P \propto Q^2$$
$$P_w \propto N^2, \quad \text{therefore } P_w \propto Q^3$$
$$\text{and } D \propto N.$$

One of these laws is useful in obtaining a series of values from the system design flow and index pd. From $P \propto Q^2$: $P_2/P_1 = (Q_2/Q_1)^2$ and $P_2 = (P_1/Q_1^2) \times Q_2^2$. Thus

$$P_2 = K \times Q_2^2$$

where $K = P_1/Q_1^2$. Using this formula allows other values of system P and Q to be generated for plotting the system characteristic although most solutions can be done without it.

Example 3.1

A heating system has design conditions of 1.5 litres/s at 50 kPa. A circulating pump having the characteristic detailed is installed. By plotting the system and pump characteristics, determine:

(a) the operating point of the pump on the system;
(b) the pd required across the regulating valve to achieve design flow;
(c) the final operating point of the pump on the system.

Pump characteristic:

P	0	20	40	60	80	100 kPa
Q	3	2.8	2.45	1.95	1.2	0 litres/s

Solution

From the derived equation above, additional points for plotting the system characteristic can be evaluated:

$$K = \frac{50}{(1.5)^2} = 22.22$$

See Table 3.1.

(a) You should now plot the pump and system characteristics. The operating point of the pump on the system where they intersect is 1.75 litres/s at 66 kPa (Figure 3.2).

Table 3.1   Example 3.1: pump and system characteristics

Q^2	Q_2^2	K	P_2
0.5	0.25	22.22	5.55
1.0	1.0	22.22	22.22
1.5	2.25	22.22	50.0
2.0	4.0	22.22	89.0

(b)   By projecting a vertical line upwards from the point of design flow (1.5 litres/s) on your graph, it intersects the pump characteristic at the final operating point of the pump on the system. Refer again to Figure 3.2. The final location of the system characteristic can now be sketched on your graph by hand. Two horizontal lines can now be drawn back to the pressure axis of your graph and the required regulating pd read from the pressure scale. This should come to $72.5 - 50 = 22.5$ kPa.

(c)   From your graph, which should now look similar to Figure 3.2, the final operating point of the pump on the system after regulation is 1.5 litres/s at 72.5 kPa.

Note: The vertical line from the design flow on your graph and in Figure 3.2 intersects two points: the original system characteristic at a pressure of 50 kPa, and the pump curve at 72.5 kPa. The first intersection is at the design pressure drop.

Example 3.2
A pump having the following characteristic is to be employed in a system whose design conditions are 3.5 litres/s at 24 kPa. Determine whether or not two identical pumps operating together are required and state whether a series or parallel arrangement should be adopted. Assess the pressure reduction required to achieve design flow.

Pump characteristic:

P	0	10	20	30	40	50	60 kPa
Q	3.0	2.85	2.6	2.28	1.85	1.3	0 litres/s

Solution
For the series arrangement the pressure developed is doubled. Refer to Figure 3.3:

$2P$	0	20	40	60	80	100	120 kPa

For the parallel arrangement the flow rate is doubled. Refer to Figure 3.4:

| $2Q$ | 6.0 | 5.7 | 5.2 | 4.56 | 3.7 | 2.6 | 0 litres/s |

By analysing the options *before* plotting the pump characteristics it is clear that:

1. a single pump is not big enough, as the maximum flow of 3.0 litres/s is achieved at zero pressure;
2. two pumps simultaneously operating in series do not increase the flow above 3.0 litres/s;
3. two pumps operating simultaneously in parallel achieve 3.7 litres/s at 40 kPa, which is in excess of the design conditions and is therefore worthy of investigation.

The solution therefore is obtained from the two pumps operating simultaneously in parallel. After obtaining further values for the system design condition of 3.5 litres/s at 24 kPa using $K = P_1/(Q_1)^2 = 1.96$, the system and *parallel* pump characteristics are plotted, and from the graph the regulation required to achieve design flow is found to be $42 - 24 = 18$ kPa. It is recommended that you now plot the characteristics and check the above solution. What is the final operating point of the pump on the system? You should have 3.5 litres/s at 42 kPa.

Example 3.3

Consider the system shown in Figure 3.9(a). Design flow rate is 2.5 litres/s and the system and pump characteristics are given in the data below. Analyse the pump pressure distribution around the system and determinewhether or not there will be discharge at the open vent. If the regulating valve is relocated to the pump discharge will the system operate satisfactorily?

Data

Hydraulic resistances around the system:

Section	A–B	B–C	C–D	D–E	E–F	F–A
Resistance (kPa)	2	2	8	2	2	3

Pump characteristic:

P	50	48	**43.5**	**32.5**	23.5	0 kPa
Q	0	1	**2**	**3**	3.5	4 litres/s

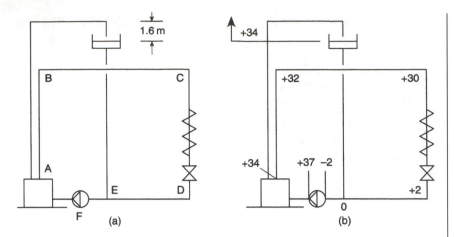

Figure 3.9   Example 3.3.

Solution

The sum of the hydraulic resistances around the system identifies the net pump pressure required, as the single circuit forms the index run. Thus the system design conditions are 2.5 litres/s at 19 kPa. Now look at the pump characteristic data and you will see that the system design conditions fall between the values in bold type, so the proposed pump is clearly big enough. From the system design conditions other values can be calculated as shown in example 3.1, and the system and pump characteristics can now be plotted and system regulation considered as shown in Figure 3.2.

The operating point of the pump on the system is 3.2 litres/s at 30 kPa. By regulating back to the design flow of 2.5 litres/s, the pump pressure developed is 39 kPa and the regulation required is 20 kPa. Knowing the pump pressure under operating conditions we can now plot the pump pressure effects around the system as shown in Figure 3.9b. Pump pressure at the pump outlet will be positive, and the pump pressure effects around the system will therefore be positive up to the neutral point at the cold feed entry, where it must change to a negative pump effect. Remember, at the neutral point, pump pressure changes. From the data the hydraulic resistance from point E to point F is 2 kPa; this will have a negative value and thus at the pump inlet the pump pressure is −2 kPa. The pressure rise through the pump is 39 kPa: thus the discharge pump pressure will be +37 kPa. By using the other section resistances from the data, the remaining pump pressure effects can now be plotted around the system. Make sure you agree with the analysis in Figure 3.9b.

Now we need to consider what happens at the open vent. If water does *not* flow in the open vent pipe the residual pump pressure effect at point A will be the same at the original water level in the open vent pipe, namely +34 kPa. This is equivalent to a rise in water level of $h = P/\rho g$ metres. Thus

$$h = \frac{34\,000}{1000 \times 9.81} = 3.46\,\text{m}$$

Clearly water *will* flow in the open vent pipe as it is only 1.6 m above the water level and some of the residual pump pressure here will be absorbed. However, it would be foolhardy to leave the system as it is: the likelihood of flow in the vent pipe is overwhelming, as the height of the pipe above the water level in the F&E tank is only 1.6 m. If water flow takes place, circulation to the system will be reduced and the tank room will fill with vapour.

How would you rectify the situation? There are two or three solutions. One of them is indicated in the question, for if the regulating valve is relocated to the pump discharge, section F–A now has a hydraulic resistance of 23 kPa and the pump effect at point A and hence at the water level in the open vent is reduced to 14 kPa, from which

$$h = \frac{14\,000}{1000 \times 9.81} = 1.43\,\text{m}$$

As the height of the vent above the water level in the tank is 1.6 m the system should operate satisfactorily.

Can you suggest other solutions to the problem of pumping over?

Example 3.4
A diagram of a system is shown in elevation in Figure 3.10(a). The design flow rate is 2.5 litres/s and the system and pump characteristics are given in the data.

(a) Determine the operating point of the pump on the system.
(b) What is the pressure loss required across the regulating valve to achieve design flow?
(c) Analyse the pump pressure distribution around the system and determine whether or not air will be drawn in through the open vent at point A.

Data
System hydraulic resistances:

Section	A–B	B–C	C–D	D–E	E–F	F–A
Resistance (kPa)	2	2	8	2	2	3

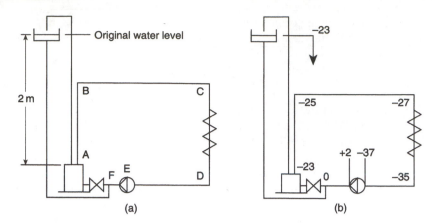

Figure 3.10   Example 3.4.

Pump characteristic:

P	50	48	43.5	32.5	23.5	0 kPa
Q	0	1	2	3	3.5	4 litres/s

Solution
You will see that the system and pump characteristics are similar to those used in example 3.3. The solution to parts (a) and (b) are therefore similar:

(a)   The operating point of the pump on the system is 3.2 litres/s at 30 kPa.

(b)   The regulating pressure drop is 20 kPa to achieve design flow of 2.5 litres/s. The pump duty is now 2.5 litres/s at 39 kPa.

(c)   Here the similarity ends, as the location of the cold feed is now at the pump discharge instead of the pump inlet. This requires another analysis of the pump effects around the system.

As the pump pressure changes from positive to negative at the neutral point F, pump discharge pressure will be +2 kPa, which from the data is the hydraulic resistance in pipe section E–F. For a pressure rise through the pump of 39 kPa the inlet pump pressure must be −37 kPa. The pump pressure effects can now be plotted anti-clockwise from pump inlet E around the system or clockwise from the cold feed entry at point F, using the hydraulic resistances for each pipe section from the data. The results are shown in Figure 3.10(b).

Using a similar argument as in example 3.3, the negative pump effect at point A is transferred to the original water level in the open vent pipe under no-flow conditions. We are reminded that the reason for this is that under no-flow conditions there is no loss in pump pressure. Thus the negative pump effect in the vent is 23 kPa, which is equivalent to

$$h = -\left(\frac{23\,000}{1000} \times 9.81\right) = -2.34\,\text{m}$$

As the F&E tank is only 2 m above point A, air will be drawn into the system at this point, and furthermore the pump will not operate in a stable manner.

Example 3.5

The diagram in Figure 3.11(a) shows a simple LTHW heating system in elevation. From the data determine:

(a) the operating point of the pump on the system;
(b) the pressure loss required across the regulating valve to achieve design flow;
(c) the antiflash margin at point E in the system if water temperature at this point is 85 °C.

Data
System characteristic:

Section	A–B	B–C	C–D	D–E	E–F	F–A
Hydraulic resistance (kPa)	5	10	10	20	20	5

System design flow: 4.0 litres/s
Pump characteristic:

P	120	114	105	95	83	67	45	0 kPa
Q	0	1	2	3	4	5	6	7 litres/s

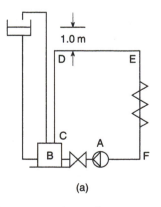

(a)

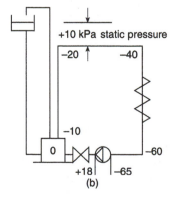

(b)

Figure 3.11   Example 3.5.

Steam tables data:

Absolute pressure (kPa)	110	100	80	70	60
Saturation temperature (°C)	102	100	93	90	86

Take atmospheric pressure as 100 kPa.

Solution

By summing up the hydraulic resistances in the index circuit, index pressure drop is 70 kPa at 4.0 litres/s. Before proceeding further, look at the pump characteristic in the data, and you will see that the nearest duty is 83 kPa at 4.0 litres/s. Clearly this pump should be suitable for the system. It also offers a flow rate that matches the system design flow. It is therefore necessary only to plot the pump and system characteristics to solve part (a) of the question.

(a)   Adopting the appropriately adapted pump law ($P_2 = K \times Q_2^2$), a series of system pressure and flow readings are obtained and the system characteristic plotted along with the pump characteristic. You should undertake this piece of work now.
      From the graph, which should be similar to Figure 3.2, the operating point of the pump on the system is 4.25 litres/s at 80 kPa.
(b)   By drawing a vertical line on your graph through the design flow of 4.0 litres/s, the required pressure drop across the regulating valve from the graph and from the system design conditions is $83 - 70 = 13$ kPa.
(c)   This solution requires an analysis of the pump pressure distribution around the system, which is shown in Figure 3.11(b).

The initial step commences at the neutral point, where the pump pressure effect passes through zero. Pump discharge pressure must therefore be $+18$ kPa to overcome the hydraulic resistance of pipe section A–B and that required of the regulating valve. For a pressure rise through the pump of 83 kPa, the inlet pump pressure must be -65 kPa. With this knowledge the pump pressure distribution around the system can now be found with the aid of the system section resistances.

At point E, the algebraic sum of pump and static pressure is $(+10 + -40)$ kPa $= -30$ kPa gauge. Clearly this is a case where sub-atmospheric pressures exist in this high-level pipe when the pump runs. Absolute pressure $P_{abs} = $ gauge $P_g + $ atmospheric $P_{at} = -30 + 100 = 70$ kPa. From the data in the question relating to the steam tables, the saturation temperature of water at 70 kPa abs is 90 °C. As the water temperature at point E is 85 °C, the antiflash margin is $90 - 85 = 5$ K. This margin is too small, a minimum margin of 10 K is required, as water becomes unstable at 5 °C below its boiling point.

Therefore unless the configuration of the pump and cold feed are changed, the maximum water temperature sustainable at point E will be $90 - 10 = 80\,°C$.

Note: Although the examples used here relate to vented systems so that pumping over and air ingress can be identified, the same procedure can be adopted for pressurized systems. In such cases, the neutral point is where the cold feed and expansion pipe from the pressurization unit enter the system.

For pressurized systems, potential problems at the open vent do not arise. However, the effects of negative pump pressure around the system should be investigated to ensure that water does not boil at any point when the system is run up to operating temperature with the pump running.

The ideal location for the pump is in the flow as shown in Figure 3.7(b). Do you agree?

3.6 Pumps having inverter speed control

Pumps with the facility to electronically vary their speed and hence flow rate, pressure developed and power consumption, according to prevailing system conditions are now in common use. This type of pump control increases the working life of the pump and reduces its consumption of electricity – two important considerations in this climate of sustainability.

A fixed speed pump is normally oversized and so a regulating valve is fitted on its discharge to reduce the flow rate to the design flow. Regulating valves like volume control dampers absorb energy and therefore indirectly consume electrical energy. This energy consumption is continuous during system operation whereas a pump (also oversized) fitted with inverter speed control will automatically reduce its speed to the design flow without that regulating valve, thus immediately and continuously saving energy during system operation.

Further reductions in energy consumption will also occur as shown in example 3.6. Clearly we are talking here about the omission of the *pump* regulating valve. Circuits that the pump is serving still require regulation for balancing purposes.

Example 3.6

A centrifugal pump has a design duty of 4 litres/s at 500 kPa when running at 2400 rpm, and an efficiency of 75%. It is fitted with electronic speed control.

(i)   Calculate the shaft power of the pump when it is operating at its design duty.

(ii) As a result of system response to solar penetration into the building the flow rate is periodically reduced to 1.5 litres/s. Calculate the shaft power when this occurs. Assume pump efficiency remains constant.

    (iii)   Express the shaft power to handle 1.5 litres/s as a percentage of that required at design duty.

    (iv)   Determine the rotational speed of the pump and the pressure developed when it delivers 1.5 litres/s.

    (v)   List the advantages of variable speed over fixed speed pumps.

Solution

    (i)   Shaft power is obtained from $P_w = (Q \times \text{pd})/\eta$ where Q is the volume flow handled by the pump in m^3/s and pd is the pressure developed by the pump in Pa. Thus $P_w = (0.004 \times 500\,000)/0.75 = 2667\,W$. You should now confirm that the product of Q and pd has the derived unit of watts.

    (ii)   From Section 3.5, $P_w \propto Q^3$ so $P_{w2} = (Q_2/Q_1)^3 \times P_{w1} = (1.5/4.0)^3 \times 2667 = 141\,W$.

    (iii) $\%P_{w1} = (P_{w2}/P_{w1}) \times 100 = (141/2667) \times 100 = 5.3\%$.

    (iv)   From Section 3.5, $Q \propto N$ so $N_2 = (Q_2/Q_1) \times N_1 = (1.5/4.0) \times 2400 = 900\,rpm$.

    (v)   Savings in electrical consumption, reduction in CO_2 emissions if the electricity is sourced from fossil fuel plant, lower noise levels when running at lower speeds, better and simpler control, longer life cycle for the pump, lower life cycle cost which takes into account the original purchase price and running costs.

3.7   Chapter closure

This completes the work on pumps and systems. You now have the necessary skills to analyse the effects of different configurations of cold feed and pump locations in closed systems and to select the best position for the pump at the design stage. You will be able to offer solutions to some of the problems associated with existing installations, such as continual air in the system, erratic operation of the pump, considerable noise in high-level pipes or from within the boiler, discharge from the open vent, and air in high-level pipes at the beginning of the heating season.

You are aware of the potential advantages of employing the variable speed pump and the contribution it can make in the debate on sustainability. Finally, you are now in a position to recommend a suitable pump type for a given application.

High-temperature hot water systems 4

A	area of surface (m²)
C	specific heat capacity (kJ/kg K)
CTVV	constant temperature variable volume
dp	pressure difference (Pa)
E	expansion volume (litres, m³)
F&E	feed and expansion
F_1, F_2	heat loss factors
h	static head (m)
HTHW	high-temperature hot water
K	radiator constant
LTHW	low-temperature hot water
m	mass (kg)
N	air changes per hour
n	natural draught convector index, thermodynamic index
P	pressure (Pa, kPa)
P_N	partial pressure of nitrogen (Pa)
P_T	total pressure (Pa)
P_{WV}	partial pressure of water vapour (Pa)
R	constant for nitrogen (J/kg K)
T	absolute temperature (K)
t_c	dry resultant, comfort temperature (°C)
t_f	flow water temperature (°C)
t_m	mean surface temperature (°C)
t_{ao}	outdoor temperature (°C)
t_r	return water temperature (°C)
t_s	saturation temperature (°C)
U	thermal transmittance coefficient (W/m² K)
V	volume (m³)
γ	thermodynamic index
ρ_1	density at fill temperature (kg/m³)
ρ_2	density at mean operating temperature (kg/m³)
$\sum$	sum of

At the beginning of Chapter 2, Table 2.1 lists operating pressures and temperatures for very-low-temperature, low-temperature, medium-temperature and high-temperature systems, which puts the subject of wet space heating systems into context.

4.1   Introduction

When water is elevated to a temperature in excess of its boiling point at atmospheric pressure (100 °C at sea level) for the purposes of space heating, the system must be pressurized to provide a minimum antiflash margin of 10 K at the weakest pressure point. This prevents the water from boiling within the system, keeping it in its liquid phase. In most cases, high-temperature hot water heating therefore requires a pressurization unit. The specification of plant and fittings must be sufficient for these to operate under the enhanced pressure imposed.

HTHW offers a higher mean water temperature than LTHW, and as the rate of heat flow is dependent upon the magnitude of the temperature difference between the heat exchanger and the room in which it is located, it follows that smaller diameter pipework can be employed to transport the heating medium for the same heat requirement.

APPLICATIONS OF HTHW

Clearly, the application of HTHW requires that the distribution is over substantial distances, to take advantage of the comparatively small pipe. Large sites, such as those for hospitals, airports and district heating, are worth consideration. The effect of using smaller distribution pipe is extended to lower costs for anchors, guides and brackets as well as pipe insulation and duct size. Weighed against this must be the use of higher-specification plant and fittings, the need for pressurization equipment, water treatment and regular water analysis, as precipitation and corrosion increase with water temperature. Temperature conversion may be required at various points on the site to bring the water temperature down to a lower level locally for radiators.

A final decision rests upon the economy offered by the distribution mains on the one hand and the additional costs on the other.

4.2   Pressurization methods

There are at least four ways in which a system may be pressurized: the use of static head, air, nitrogen or steam. The most common is the use of nitrogen. Occasionally available static head is used for upgrading an existing installation. The use of nitrogen and static head will be considered here. There are a number of manufacturers of pressurization equipment, which is delivered to site in a packaged format. You should familiarize yourself with manufacturers' details.

TEMPERATURE CONVERSION

If radiators are used as the space heating appliances, local conversion from high to low temperature may be required. Ideally, appliances such as natural draught convectors, fan coil units, unit heaters and radiant tube or strip allow direct use of the high-temperature water, thus avoiding the cost of conversion plant.

There are two methods for space heating by which conversion is achieved, and these are shown in Figures 4.1 and 4.2. Figure 4.1 illustrates the use of the injector valve. The primary circuit is the HTHW distribution and the secondary circuit is the LTHW system. Note that the secondary circuit is subject to the full HTHW pressure. If connection A–B is near the primary pump, dp A–B will be high. Without balance pipe 1, flow reversal can occur in pipe 2, as the secondary LTHW pump will develop a lower pressure than the primary pump. Pipe 3 mixes LTHW return with the flow at low load when there is no flow in pipe 2. When the LTHW load increases, water from the return also circulates through pipe 2, with the injector valve providing constant-temperature water to the secondary circuit, which may require further temperature control. The valves shown are regulating except for the valve on the primary flow balance pipe, which is for isolating purposes.

Figure 4.2 shows temperature *and* pressure conversion using a non-storage heating calorifier, which effectively takes the place of the boiler. The three-way

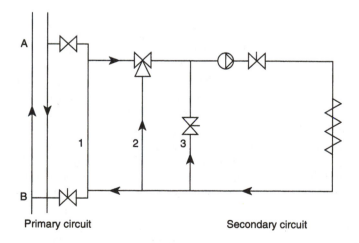

Primary circuit   Secondary circuit

Figure 4.1 Use of the injector valve for conversion from high to low temperature.

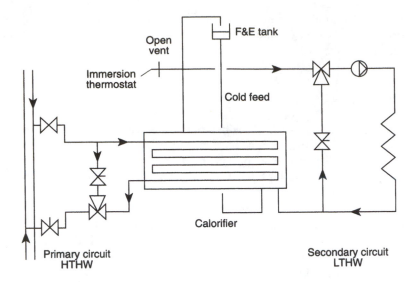

Figure 4.2 Use of the non-storage calorifier for conversion from high to low temperature and pressure.

mixing valve on the HTHW primary circuit controls the LTHW secondary flow temperature at a constant value within the immersion thermostat's differential. A three-way mixing valve is shown on the secondary circuit to provide compensated control via an outdoor detector. You will see that the secondary system is filled and vented via a local F&E tank.

PRESSURIZATION BY STATIC HEAD

Where a higher output is required from an existing LTHW system it may be convenient to utilize the static head imposed by the F&E tank to raise the boiler flow temperature. Figure 4.3 shows the conversion. Note the expansion vessel required adjacent to the boiler. It should be sized to accept all the expansion water from the system, to prevent high-temperature water reaching the F&E tank. The tank accepts the cold water displaced from the expansion vessel on heat-up. The expansion vessel is fitted with sparge pipes to inhibit mixing of the contents during heat-up.

Clearly, the value of static head h needs to be substantial in order to raise boiler flow temperature much above 90 °C, to prevent the water boiling in the high-level pipes (usually the weakest point in the system). One metre of static height h is equivalent to 10 kPa. Five metres is equivalent to half an atmosphere. Recourse to the saturation temperatures in the steam tables shows at 1.5 bar absolute (0.5 bar gauge) a value of 112 °C. With an antiflash margin of 12 K the boiler flow temperature can be raised to 100 °C.

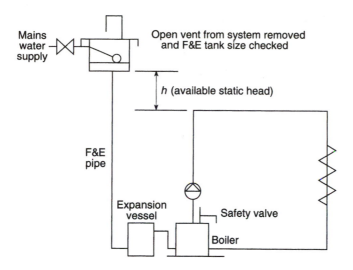

Figure 4.3   Pressurization by static head.

Raising the boiler flow temperature can raise the mean water temperature and hence the system output as long as the appliances can operate at 100 °C. We are also assuming here that the boiler has sufficient capacity to service the additional output, otherwise that will need upgrading.

This scenario does not account for the effects of pump pressure around the system, which, when the pump operates, is subject to the algebraic sum of the static and pump pressures at any given point. Chapter 3 shows clearly that negative pump pressure can have an adverse effect upon the combined pressure at a point in the system – particularly along high level pipes. The combined effect of pump and static pressure needs to be used in an analysis of this kind before enhancing the system flow temperature.

Expansion volume E for sizing the vessel is obtained from:

$$E = V \left(\frac{\rho_1 - \rho_2}{\rho_2} \right) \tag{4.1}$$

where suffix 1 = initial or fill temperature (°C); suffix 2 = mean water temperature (°C); and V = system contents (litres, 1, or m^3).

CONSEQUENCES OF UPGRADING

If limited data are available in a situation where a building's space heating system is to be upgraded, a heat balance may be drawn with building heat loss and heat emission from the terminals.

From Section 1.2 in Chapter 1, for steady-state temperatures, heat loss = heat emission. Thus:

$$\left(\sum (UA)F_1 + C_v F_2\right)(t_c - t_o) = KA(t_m - t_c)^n$$

Ignoring the constants:

$$(t_c - t_o) \propto (t_m - t_c t)^n$$

Therefore

$$\frac{(t_c - t_{ao})_2}{(t_c - t_{ao})_1} = \frac{(t_m - t_c)_2^n}{(t_m - t_c)_1^n} \qquad (4.2)$$

An example of a system upgrade is

Example 4.1

If an LTHW heating system operating on natural draught convectors is to be upgraded, determine the probable indoor temperature achieved given that the original design conditions were: room $t_c = 14\,°C$, outdoor $t_o = -1\,°C$, system flow $t_f = 80\,°C$, and system return $t_r = 70\,°C$, and the proposed new design conditions are: $t_o = -1\,°C$, $t_f = 100\,°C$ and $t_r = 90\,°C$. The index for natural draught convectors $n = 1.5$.

Solution

For the original design the difference between mean water and comfort temperature is $75 - 14 = 61\,K$. After upgrading, the temperature difference is greater in value but the new comfort temperature will rise above $14\,°C$.

Substituting the known data into equation (4.2):

$$\frac{t_c + 1}{14 + 1} = \left(\frac{95 - t_c}{75 - 14}\right)^{1.5}$$

from which

$$\frac{t_c + 1}{15} = \left(\frac{95 - t_c}{61}\right)^{1.5}$$

By giving values to t_c of say $18\,°C$, $20\,°C$, $22\,°C$ and $24\,°C$, heat loss and heat emission can be evaluated (Table 4.1) and the results plotted to find probable comfort temperature. The plot is shown in Figure 4.4. From the plot you will see that the heat loss and convector emission has increased by 37% while comfort temperature has risen from $14\,°C$ to $19.7\,°C$.

Table 4.1   Example 4.1

Comfort temperature (°C)	Heat loss, $(t_c + 1)/15$	Heat emission, $((95 - t_c)/61)^{1.5}$
18	1.267	1.418
20	1.4	1.363
22	1.53	1.309
24	1.67	1.256

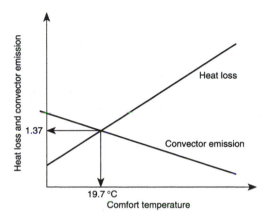

Figure 4.4   The plot for example 4.1.

Summarizing:

Condition	t_m	t_c	$(t_m - t_c)$
Design	75	14	61
Upgrade	95	19.7	75.3

The effect of the upgrading is an increase in comfort temperature and an increase in system output to offset the corresponding increase in heat loss. Since the system output is increased by 37% the mass flow rate will increase and this will necessitate an upgrade to the pump that will also be influenced by the increased pressure loss through the index pipework and fittings.

PRESSURIZATION BY GAS

Nitrogen is normally used as the gas cushion because it is inert and therefore does not react with water at its interface in the expansion vessel. Small systems that are pressurized use expansion vessels fitted with a membrane to divide the water and gas cushion. Because of the low static head available for rooftop plant rooms, a pressurization unit would normally be used without necessarily elevating the flow temperature above 85 °C.

Some specifications call for the space heating system to be pressurized as a matter of course, as this dispenses with the problems associated with the remoteness of the F&E tank and ball valve and the 'out of sight, out of mind' syndrome during maintenance. The pressurization unit, however, is always located close to the boiler plant.

DOMESTIC PRESSURIZATION

Figure 4.5 shows a domestic heating system that is pressurized and connected directly to the mains water service.

Where direct connection to the rising main has not been made, a hose union connection is an alternative. A regular check on the pressure gauge is necessary, however, so that topping up can be undertaken manually.

COMMERCIAL PRESSURIZATION

Pressurization units come as a package, which includes duty feed pump and standby, expansion vessel, feed tank, pressure switches and pressure gauge. It is only a matter of locating the unit adjacent to the boiler plant, and connecting the feed pipe from the unit to the boiler or system return and the mains water service to the ball valve connection on the feed tank.

The expansion vessel and pump need to be sized and the high and low pressures determined for selecting the pressure switches. These matters relate to the system to which the pressurization unit is to be connected, and depend upon the selected design flow and return temperatures, water content and

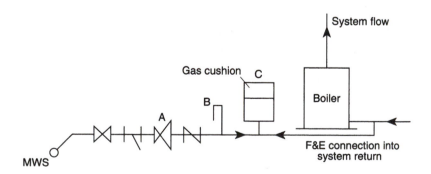

Figure 4.5   Domestic pressurization off the rising main: A = pressure-limiting valve; B = relief valve; C = expansion vessel.

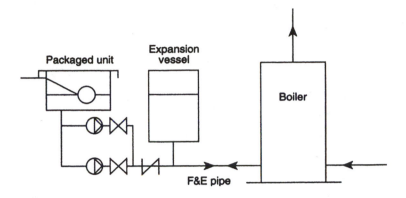

Figure 4.6 Commercial pressurization.

the combined effects of the static pressure and pump pressure around the system.

A typical arrangement is shown in Figure 4.6.

THE NEUTRAL POINT

It is worth a reminder that the neutral point occurs at the entry of the F&E pipe into the system. This is important when considering the combined effects of system pump and static pressures.

4.3   Pressurization of large systems

Practice in the UK adopts the pressurization unit with the spill tank, whereas in the rest of Europe the expansion vessel is sized to accept all the system expansion water, and therefore the unit is similar to that shown in Figure 4.6. The expansion vessel can become quite large as a result. To prevent this happening, and to avoid the costs of the expansion vessel, which is classed for insurance purposes as a pressure vessel, the spill tank takes most of the system expansion water, thus making the expansion vessel relatively small in size. Figure 4.7 shows this type of unit.

As the system water is heated, the expansion water enters the F&E pipe and passes into the expansion vessel. The high-pressure controller eventually activates the spill valve, and further expansion water is diverted to the spill tank until system operating temperature is reached. Excess pressure is relieved after initial start-up by using the hose union valve and running unwanted water to drain. When the system load fluctuates, the quantity of expansion water varies. On a fall in system heating load, the system water contracts, resulting in a fall-off in pressure. The low-pressure controller is activated, and the feed pump

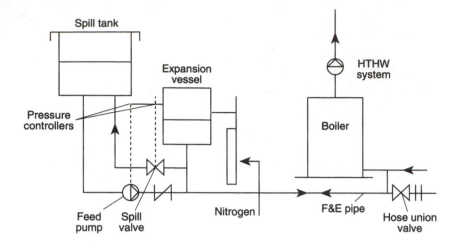

Figure 4.7   Pressurization unit with spill tank.

draws water from the spill tank to replenish the system. The pressure difference between the high-pressure and low-pressure controllers must be significant – usually about 1 atm – to ensure that they operate correctly.

SIZING THE PRESSURIZATION UNIT

This would normally be done by the manufacturer of the pressurization unit. However, you will have to provide the system specification so that the manufacturer can select the appropriate packaged unit. It is therefore useful to have some knowledge of how selection is made.

System filling and operating cycle

If the filling process takes place quickly with the expansion vessel insulated, it will be close to adiabatic, and $PV^\gamma = C$. If the filling process takes place slowly with the vessel uninsulated, it will be close to isothermal and $PV = C$. Expansion vessels are always left uninsulated, and the time taken to fill is controlled: so the process is polytropic, where $PV^n = C$. The indices $\gamma > n > 1$, where index n for the polytropic process is taken as being 1.26.

To size the expansion vessel one of the above laws is adopted. During the filling process there are three stages:

1. system filled – unpressurized;
2. system filled – pressurized and at fill temperature;
3. system pressurized and running at operating temperature.

Consider the three stages in Figure 4.8.

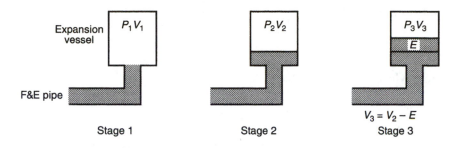

Figure 4.8 The three stages in system pressurization.

SIZING THE EXPANSION VESSEL

Conveniently, V_1 in Figure 4.8 is vessel volume or size as well as the initial gas volume. The other gas volumes are unknown, but the pressure P at each stage is known or can be calculated. A formula needs to be derived, therefore, for determination of vessel size in terms of the stage pressures and the expansion volume E for the system, which can be calculated from equation (4.1).

Using Boyle's law, $PV = C$, assuming initially, isothermal conditions. Now

$$P_2 V_2 = P_3 V_3 = P_3 (V_2 - E)$$

Then

$$P_2 V_2 = P_3 V_2 - P_3 E$$

and

$$P_3 E = P_3 V_2 - P_2 V_2$$

So

$$P_3 E = V_2 (P_3 - P_2)$$

from which

$$V_2 = \frac{P_3 E}{P_3 - P_2}$$

Now

$$P_1 V_1 = P_2 V_2$$

Substituting for V_2:

$$P_1 V_1 = \frac{P_2(P_3 E)}{P_3 - P_2}$$

from which

$$V_1 = \left(\frac{P_2}{P_1}\right)\left(\frac{P_3 E}{P_3 - P_2}\right) \tag{4.3}$$

This is the volume of gas at stage 1 in the pressurization process. It also conveniently is the size of the expansion vessel.

Thus the expansion vessel can be sized from a knowledge of stage pressures and the system expansion volume.

The equation is found in some texts in the following format:

$$V_1 = \left(\frac{P_2}{P_1}\right)\left(\frac{E}{1 - (P_2/P_3)}\right) \tag{4.4}$$

Can you reconcile this formula with the derived equation (4.3)?

It is now a simple matter to amend the formula for a polytropic process by introducing index n:

$$V_1 = \left(\frac{P_2}{P_1}\right)^{1/n}\left(\frac{E}{1 - (P_2/P_3)^{1/n}}\right) \tag{4.5}$$

Can you reconcile this formula with equation (4.4)? Look closely at the derivation of equation (4.3).

EFFECTS OF ALTITUDE

Gravitational acceleration g is taken as $9.81\,\mathrm{m/s^2}$ at sea level. Clearly this is reduced in value with the effects of altitude and therefore should be borne in mind in the design of systems in locations much above sea level.

Example 4.2
Figure 4.9 shows a simple HTHW heating system in elevation. The system data are as follows.

Data

Proposed boiler flow temperature	165 °C
Filling pressure	1.5 bar gauge
System contents	2200 litres
Density on filling	1000 kg/m³
Density at mean water operating temperature	890 kg/m³
Antiflash margin	15 K

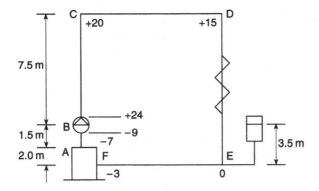

Figure 4.9 Example 4.2: HTHW system pressurized by gas with the pump pressure effects added.

Section	A–B	B–C	C–D	D–E	E–F	F–A
Hydraulic resistance (kPa)	2	4	5	15	3	4

You are asked to check the antiflash margin at the weakest point in the system, determine the minimum operating pressure in the expansion vessel, size the expansion vessel and calculate the mass of nitrogen required for the gas cushion. Assume the pump is operating and then consider the effects of pump failure.

Solution
The hydraulic resistances around the index circuit total 33 kPa, and with the neutral point in the system at E the pump pressure effects can be added to the diagram as shown in Figure 4.9. If you have any difficulty here go back to Chapter 3 and look again at the examples.

The static heights need to be converted to static pressures working from point A:

$$1.5\,\text{m to point B is equivalent to } 1.5 \times 890 \times 9.81 = 13\,\text{kPa}$$

$$9.0\,\text{m to points C–D is equivalent to } 9.0 \times 9.81 = 79\,\text{kPa}$$

Point E is 2.0 m below point A. This is equivalent to

$$2.0 \times 890 \times 9.81 = 17\,\text{kPa}$$

Datum pressure
Point A in Figure 4.9 will be considered the datum, and datum pressure here will need to support with safety the proposed boiler flow temperature of 165 °C. Applying the antiflash margin of 15 K, the supporting pressure will

correspond to a saturation temperature of $165 + 15 = 180\,°C$, and from the steam tables this is 10 bar absolute, 9.0 bar gauge.

This *includes* the static pressure imposed by the height of the system above point A. Static, pump and combined pressures can now be tabulated around the system (Table 4.2).

The weakest point in the system
The combined pressures are now considered for pump running conditions. However, these pressures are about the datum at point A of 10 bar absolute, 9 bar gauge. The weakest point in the system is at D, where the combined static and pump pressure is $-64\,kPa$ or $-0.64\,bar$. The gauge pressure at this point, taking datum pressure into account, will therefore be: datum minus the combined pressure at point D. That is:

$$9 - 0.64 = 8.36 \text{ bar gauge}$$

From the steam tables, the saturation temperature at this pressure is approximately $177\,°C$. Flow temperature at point D is likely to be close to the boiler flow of $165\,°C$, so the antiflash margin at this point is $(177 - 165) = 12\,K$. The minimum acceptable antiflash margin is $10\,K$, so the proposed boiler flow temperature for this system is acceptable.

Operating pressure in the expansion vessel, P_3
The operating pressure in the expansion vessel can also be determined from the tabulated analysis. At point E the combined pressure is $+17\,kPa$, so gauge pressure here will be datum plus the pressure at point E, namely $9 + 0.17 = 9.17\,bar$ gauge. However, the water level in the expansion vessel is 3.5 m above this point. This is equivalent to a loss of $30\,kPa$ static pressure. Thus the pressure at the water level will be:

$$9.17 - 0.3 = 8.87 \text{ bar gauge, } 9.87 \text{ bar absolute}$$

This is the *operating pressure*, P_3, in the expansion vessel.

Table 4.2   Example 4.2: pressures round the system (kPa)

Point	Static	Pump	Combined
A	0	−7	−7
B inlet	−13	−9	−22
B outlet	−13	+24	+11
C	−79	+20	−59
D	−79	+15	**−64**
E	+17	0	+17
F	+17	−3	+14

If it is essential to maintain an antiflash margin of 15 K at point D with the pump running, boiler pressure (i.e. *datum* pressure) would have to be increased from 9 bar gauge to 9.64 bar gauge, and the resultant pressure at the water level in the expansion vessel,

$$P_3 = revised\ datum + 0.17 - 0.3 = 9.64 + 0.17 - 0.3$$

$$= 9.51\ \text{bar gauge},\ 10.51\ \text{bar absolute}$$

Sizing the expansion vessel

Initial pressure P_1 can be taken as 1 atm, 1.0 bar absolute. Filling pressure P_2 is given as 1.5 bar gauge, 2.5 bar absolute. Expansion volume E is obtained from equation (4.1):

$$E = 2200 \left(\frac{1000 - 890}{890} \right) = 272\ \text{litres}$$

Substituting the stage pressures and the system expansion volume for a 15 K antiflash margin at point D, adopting equation (4.3) for isothermal conditions:

$$V_1 = \frac{2.5}{1.0} \times \frac{10.51 \times 272}{10.51 - 2.5} = 893\ \text{litres}$$

For a 12 K antiflash margin at point D:

$$V_1 = \frac{2.5}{1.0} \times \frac{9.87 \times 272}{9.87 - 2.5} = 911\ \text{litres}$$

Mass of nitrogen

If the vessel accepts all the expansion volume, in this case 272 litres, the mass of nitrogen m_3 can be determined from the *characteristic gas equation*, which states that

$$PV = mRT$$

Now, as $P_1 V_1 = P_3 V_3 = m_3 R T_3$:

$$m_3 = \frac{P_1 V_1}{R T_3}$$

If water/gas temperature within the vessel is constant at 20 °C, $T_3 = 293$ K. P_1 is taken as 1 atm, and consists of the sum of the partial pressures of the water vapour and the nitrogen. At 20 °C the partial pressure of the water vapour is negligible. You should check this fact in the steam tables. Thus $P_1 = P_N = 1.0$ bar absolute $= 100\,000$ Pa. The gas constant R for nitrogen $= 297$ J/kg K, and taking V_1 as 893 litres, which is $0.893\,\mathrm{m}^3$, the mass of nitrogen forming the gas cushion will be

$$m_3 = \frac{100\,000 \times 0.893}{297 \times 293} = 1.026\,\mathrm{kg}$$

If you are not familiar with the characteristic gas equation you should note the units of the terms in the formula.

Pump failure
If the pump fails, the system reverts to static head conditions, and the weakest points occur at the lowest static pressure, which is along $C - D$. From the tabulated analysis the pressure here will be datum minus 79 kPa. Operating pressure along $C - D = 9 - 0.79 = 8.21$ bar gauge, 9.21 bar abs. From the steam tables, saturation temperature at this pressure is about 177 °C. Water temperature along $C - D$ will be approximately the same as the boiler flow temperature, namely 165 °C. Thus the antiflash margin $= (177 - 165) = 12\,\mathrm{K}$, which is within the safety limit.

The polytropic process
This most nearly equates with practical system pressurization. Substituting the stage pressures for a minimum antiflash margin of 15 K throughout the system and expansion volume E into the polytropic equation (4.5) for the expansion vessel volume V_1,

$$V_1 = \left(\frac{2.5}{1.0}\right)^{1/1.26} \times \frac{272}{1 - (2.5/10.51)^{1/1.26}} = 827\,\mathrm{litres}$$

The vessel volume is a little less than that for the isothermal process.

Use of the spill tank
This example assumes that all the expansion water from the system enters the expansion vessel. If the pressurization unit includes the feature of a spill tank open to atmosphere, the expansion vessel is much reduced in size and cost.

Thus when, say, 75% of the system expansion water is diverted to the spill tank, the polytropic process yields the following solutions:

$$\text{spill tank size} = 827 \times 0.75 = 620\,\mathrm{litres}$$

$$\text{expansion vessel size} = 827 \times 0.25 = 207\,\mathrm{litres}$$

The results are summarized in Table 4.3.

Table 4.3 Example 4.2: summary of results

Pump	Antiflash margin (K)	Weakest point in system	Pressure at weakest point (bar gauge)	Minimum operating pressure in expansion vessel (bar gauge)	Size of expansion vessel (litres)	Mass of nitrogen (kg)	Process
On	12 (15)	D	8.36	8.87 (9.51)	911 (893)	1.026	Isothermal
Off	12	C – D	8.21				Isothermal
On	15				827		Polytropic

DALTON'S LAW OF PARTIAL PRESSURES

The total pressure within the vessel at any operating point is the sum of the partial pressures that include the partial pressure exerted by the water vapour and the partial pressure required of the nitrogen. Thus $P_T = P_{WV} + P_N$.

Example 4.3

An expansion vessel has a 100 litres nitrogen cushion. If the water is initially at a temperature of 130 °C and the operating pressure P_T in the vessel under these conditions is 10 bar absolute, determine:

(a) the partial pressure of the nitrogen;
(b) the mass of nitrogen forming the gas cushion.

Solution

(a)   Dalton's law of partial pressures in this context is $P_T = P_N + P_{WV}$. From the steam tables the partial pressure exerted by the water vapour at a temperature of 130 °C is 2.7 bar absolute. Thus the partial pressure of the nitrogen will be $(10 - 2.7) = 7.3$ bar absolute.

   (b)   From example 4.2, $m = PV/RT$ kg. Selecting the correct units before substitution:

$$m = \frac{730\,000 \times 0.1}{297 \times (273 + 130)} = 0.61\,\text{kg}$$

Clearly, when the system reverts to ambient temperature there will be a fall-off in pressure within the expansion vessel due to the drop in the partial pressure of the water vapour. Either more nitrogen is required in the vessel or the feed pump will be energized to re-establish the operating pressure in the vessel.

At ambient conditions the partial pressure of the water vapour is insignificant and ignored. Thus $P_T = P_N$.

Example 4.4
Figure 4.10 shows in elevation a diagrammatic view of an HTHW system pressurized by gas serving an air heater battery. The required flow temperature at the heater battery is 150 °C with a temperature drop across the battery of 20 K. An antiflash margin of 15 K is required at all points in the system under all operating conditions when the pump is running.

From the data, analyse the performance of the system and offer comment. Propose a minimum acceptable boiler operating pressure.

Data
Take water density as 918 kg/m³.

Pipe section	A–B	B–C	C–D	D–E	E–A
Hydraulic resistance (kPa)	5	109	5	5	15

Solution and analysis
Converting the static heads to pressures:

for 1 m, $P = 1 \times 918 \times 9.81 = 9 \, \text{kPa}$
for 12 m, $P = 12 \times 918 \times 9.81 = 108 \, \text{kPa}$

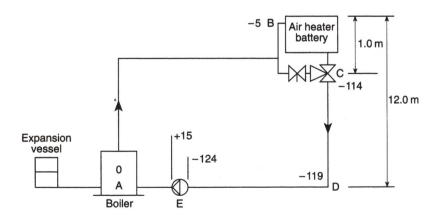

Figure 4.10   Example 4.4: HTHW system pressurized by gas with pump pressure effects added.

Table 4.4   Example 4.4: pressure effects in the system

Point	Static	Pump	Combined
A	0	0	0
B	−108	−5	−113
C	−99	−114	**−213**
D	0	−119	−119
E_{in}	0	−124	−124
E_{out}	0	+15	+15

The net pump pressure developed will be the sum of the index hydraulic losses = 139 kPa.

Ignoring heat loss from the flow main to the battery, the boiler flow temperature will be 150 °C. Using the antiflash margin here of 15 K, the supporting pressure corresponding to a saturation temperature of $(150 + 15) = 165$ °C is 7 bar absolute. Taking this initially as datum at point A in the diagram, the static, pump and combined pressure effects can now be tabulated (Table 4.4).

The weakest point in the system is at C, where the pressure = $7 − 2.13 = 4.87$ bar abs, and from the steam tables saturation temperature $t_s = 151$ °C.

When the battery load is not required, and this will occur on maximum rise in entering air temperature into the battery, the CTVV control will ensure that the system water bypasses the heater, and its temperature at point C will be the same as it is in the flow main, namely, 150 °C. Saturation temperature at this point is 151 °C, and serious cavitation will occur. Point D and point E at the pump inlet will also be reduced to approximately 8 K when the system is on full bypass. This is below the minimum acceptable value.

When the heater is operating at design load the system water temperature at point B will be 150 °C, and the antiflash margin at this point is about 8 K. This is below the minimum acceptable antiflash margin.

Check that you agree with the comments at points B, D, and pump inlet E.

Summary and way forward

Clearly, the supporting pressure selected initially was too low and needs to be increased. Taking the weakest point in the system at C and working on the maximum temperature that will occur at this point when the battery is on full bypass, namely 150 °C, and applying the minimum antiflash margin of 15 K means that the supporting pressure here must be equivalent to the saturation temperature of $(150 + 15) = 165$ °C. This is 7.0 bar absolute. Thus the pressure at the boiler to maintain this pressure at point C when the pump operates must be $(7.0 + 2.13) = 9.13$ bar absolute, 8.13 bar gauge. Refer to point C in Table 4.4 for confirmation. This is the minimum operating pressure for the boiler.

Check the antiflash margin around the system when the battery is under design load. Do you agree that the minimum operating pressure for the boiler is still 8.13 bar gauge? What effect does the location of the pump in the flow pipe have on the weakest point in the system?

With further data relating to the system it would then be a simple matter to size the expansion vessel.

4.4   Roof-top plant rooms

Boiler plant rooms located at roof level have very little static head available to maintain an adequate pressure within the boiler. Since maximum water temperature is generated within the boiler it is essential to have an adequate antiflash margin here. For this reason a pressurization unit is always used even when the system is operating on LTHW at 85 °C. There now follows an example of an MTHW system with roof-top plant.

Example 4.5
An MTHW system serving ceiling mounted fan coil units is shown in Figure 4.11. The required boiler flow temperature is 115 °C with a design temperature drop of 15 K. A minimum antiflash margin of 15 K is required at all points around the system during normal operation with the pump running. Each fan coil unit is controlled by a motorized three-port valve responding to a room thermostat.

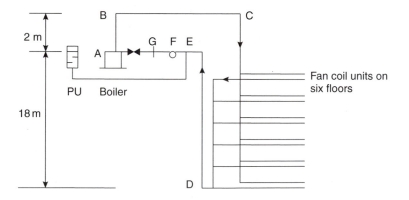

Figure 4.11   Showing the system in elevation for example 4.5 where PU is the pressurization unit and point F is the location of the pump.

Data
Water density at operating temperature is 947 kg/m³, system water content is estimated at 1800 litres, filling pressure is 30 kPa.

Section	A–B	B–C	C–D	D–E	E–F	F–G	G–A	R/valve
Hydraulic resistance (kPa)	5	5	80	15	–	–	25	20

(i)   On a sketch in elevation analyse the pump pressure effects around the system and then tabulate the static, pump and combined pressure effects.
(ii)   Locate the weakest point and find the antiflash margin when all the fan coil units are on full load and again when they are all on full bypass.
(iii)   Confirm the required boiler pressure to ensure that the minimum antiflash margin is maintained at all points around the system during normal operating conditions.
(iv)   Size the expansion vessel.
(v)   Repeat tasks (i), (ii) and (iii) when the feed and expansion pipe is connected at point G on the pump discharge.
(vi)   Give two reasons why the pump should be located in the boiler return with the F&E pipe connected on the pump suction.

Note the difference between the temperature control of fan coil units here and the control of fan convectors where the room thermostat controls the convector fan.

Solution
(i)   Figure 4.12 shows the pump pressure effects around the system in which the neutral point is at E and the pump pressure developed adds up to 150 kPa. Converting the static heads to pressures using the mean density quoted: $P_{2m} = 19$ kPa and $P_{18m} = 167$ kPa. The datum supporting pressure at the boiler corresponding to a saturation temperature of $(115 + 15) = 130\,°C$ is 2.7 bar absolute at point A. Table 4.5 shows the static, pump and combined pressures around the system.
   (ii)   From the table the weakest points occur at E and F_{in} and $P =$ datum $- 0 = 2.7$ bar abs. At this pressure saturation temperature $t_s = 130\,°C$. On full load the system return temperature at points E and F_{in} will be $115 - 15 = 100\,°C$ and the antiflash margin will be $130 - 100 = 30$ K.
   When the system fan coils are on full bypass return water temperature at points E and F_{in} will be $130 - 15 = 115\,°C$ and the antiflash margin will be $130 - 115 = 15$ K.
   (iii)   The supporting pressure required at the boiler with the pump operating will therefore be 2.7 bar abs, 1.7 bar gauge.

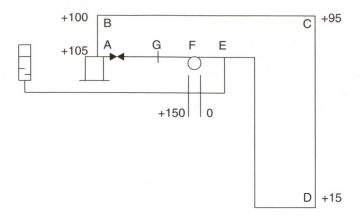

Figure 4.12   Showing the pump pressure distribution for parts (i), (ii) and (iii) of example 4.5.

Table 4.5   Showing tabulated results for example 4.5 part (i)

Point	Static	Pump	Combined
A	0	+105	+105
B	−19	+100	+81
C	−19	+95	+76
D	+167	+15	+182
E	0	0	0
F_{in}	0	0	0
F_{out}	0	+150	+150
G	0	+150	+150

(iv)   Expansion volume $= V(\rho_1 - \rho_2)/\rho_2 = 1800(1000 - 947)/947 = 1011$.

(v)   Figure 4.13 shows the pump pressure effects around the system with the neutral point now at G. The static, pump and combined pressure effects can now be tabulated about the datum boiler pressure of 2.7 bar abs. Refer to Table 4.6.

From the tabulation the weakest points when the pump is operating are at E and F_{in} where $P =$ datum $- 150\,\text{kPa} = 2.7 - 1.5 = 1.2\,\text{bar abs}$. At this pressure saturation temperature $t_s = 105\,°\text{C}$ and on full load return temperature at E and F_{in} will be $115 - 15 = 100\,°\text{C}$. So the antiflash margin will be $105 - 100 = 5\,\text{K}$. This is clearly too small for safe operation since the water will be close to boiling at these points.

When the fan coil units are on full bypass, return water temperature at point E and F_{in} will be $115\,°\text{C}$ and with the saturation temperature at

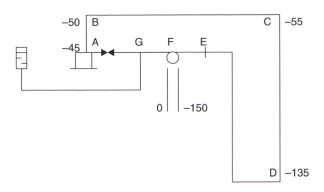

Figure 4.13   Showing the pump pressure distribution for part (v) of example 4.5.

Table 4.6   Showing tabulated results for example 4.5 part (v)

Point	Static	Pump	Combined
A	0	−45	−45
B	−19	−50	−69
C	−19	−55	−74
D	+167	−135	+32
E	0	−150	−150
F$_{in}$	0	−150	−150
F$_{out}$	0	0	0
G	0	0	0

105 °C the water will already be boiling causing serious danger. To avoid this the boiler pressure would need to be increased to $2.7 + 1.5 = 4.2$ bar abs, 3.2 bar gauge. However it is not good practice to have the boiler and system operating under negative pump pressure particularly when the static head on the boiler is so low. Refer back to Chapter 3.

(vi)   With the pump and F&E connection as shown in Figure 4.11, the boiler is under positive pump pressure ensuring a good antiflash margin at the point in the system of highest temperature. The system is also under positive pump pressure ensuring against sub-atmospheric pressures in high level pipes B–C when the pump is operating. With the F&E connection at point G and the boiler operating at 3.2 bar gauge there is a serious risk of sub-atmospheric pressures at A and along B–C.

PUMP REGULATION

As is usual practice the pump is oversized to ensure successful commissioning. As the pump in example 4.5 is a fixed speed unit, the final regulation is done at the regulating valve on the pump discharge. If a variable speed pump is used it would be oversized in the usual way but the pump regulating valve would not be required as the pump would adjust its speed of rotation downwards to match the design flow rate. The effect is to increase the life of the unit and reduce energy consumption and CO_2 emissions.

4.5   Further reading

Building services OPUS Design File (pressurization units).

4.6   Chapter closure

This completes the work on high-temperature hot water systems. You can now advise on the suitable applications for HTHW and on its merits and demerits. You now have the skills necessary to check a proposed boiler flow temperature and supporting pressure, to ensure an adequate antiflash margin around the system when the pump is operating and on pump failure, to select and size the pressurization unit, to consider the possibility of upgrading an existing system and recommend the necessary alterations. You will also be able to advise on the methods of temperature conversion and identify when it is necessary.

Steam systems 5

A	surface area (m^2)
d	diameter of pipe (m)
dh	difference in enthalpy (kJ/kg)
dp	pressure difference (Pa, kPa)
dt	temperature drop (K)
dZ	$Z_1 - Z_2$
h_f	sensible heat in condensate (kJ/kg)
h_{fg}	latent heat of saturated steam (kJ/kg)
h_g	enthalpy of saturated steam (kJ/kg)
h_w	heat in wet steam (kJ/kg)
HP	high pressure
HWS	hot water service
k	velocity pressure loss factor
k_t	total velocity pressure loss factor
K_6	velocity factor
l_e	equivalent length of pipe when $k = 1.0$ (m)
L	length of pipe (m)
LP	low pressure
LTHW	low-temperature hot water
M	mass flow rate (kg/s)
M_f	mass flow of flash steam (kg/s)
P	pressure (Pa, kPa, bar)
pd	pressure difference (Pa, kPa)
PRV	pressure-reducing valve
q	dryness fraction
Q	heat output (kW)
TEL	total equivalent length of straight pipe (m)
u	velocity (m/s)
v	specific volume (m^3/kg)
VFR	volume flow rate (m^3/s)
V_g	specific volume of the saturated vapour (m^3/kg)
Z_1, Z_2	initial and final pressure factors

5.1 Introduction

You are strongly advised to obtain manufacturers' literature relating to the application of steam and condensate equipment and fittings. Boiler manufacturers will provide literature for steam generators [*CIBSE Building Services OPUS Design File*]. Spirax Sarco Ltd., Cheltenham will provide literature relating to steam fittings and the handling of condensate and also learning material for interested students.

As a general rule, steam used for space heating is considered only when it is required also for other uses, as in manufacturing processes, for sterilization or as exhaust steam from a turbine. It is rarely generated as a discrete space heating medium. The applications for steam space heating are therefore limited to factories, chemical plants, hospitals, steam plants for the generation of electricity etc. The reason for this lies mainly in the fact that steam systems, for the purposes of economy, must normally return the condensate for reuse, and herein lies one of the disadvantages. Other disadvantages include plant response to fluctuating loads and the length of the start-up period from cold to operating temperature, and hence its inappropriate use for intermittent operation. There is also a need for water treatment and continuous analysis.

Its main advantage lies in the heat content of the latent heat of evaporation, which is used in the heat transfer process at the heat exchanger, and the high surface temperature, making the exchanger surface smaller than its LTHW counterpart. The rate of mass flow and distribution pipe sizes are different as well.

The mass flow rate of LTHW to serve a terminal whose output is 100 kW is 2.38 kg/s, which will require 50 mm flow and return distribution pipes. The corresponding mass flow of steam at 2.0 bar gauge 0.9 dry is 0.051 kg/s, which will require a 32 mm distribution pipe and a 20 mm condense return. Thus steam as a heat distribution medium is well qualified.

You will need access to the *Thermodynamic and Transport Properties of Fluids* by Mayhew & Rogers, commonly called the steam tables. Access will also be required to the *CIBSE Guide C* (Section C4) for steam pipe sizing tables. Table 5.1 included here gives a selected extract of the tables for use with examples 5.2, 5.4 and 5.5.

5.2 Steam systems

There are a number of different types of steam systems. They include systems working under vacuum and pressurized conditions. Exhaust steam from turbines and process is used after cleaning to remove oil. Steam distribution with condensate returned separately is normal practice. However, there are a few older systems in which the steam and condensate are carried in the same pipe.

Table 5.1 Selected extract from the *CIBSE* pipe-sizing tables for steam for use with examples 5.2, 5.4, 5.5 and 5.8

dZ/L	Pipe diameters									
	32 mm		40 mm		50 mm		65 mm		90 mm	
	M	l_e	M	l_e	M	l_e	M	l_e	M	l_e
90									0.402	4.8
100			0.048	1.6			0.185	3.2		
200	0.045	1.4	0.069	1.7	0.131	2.4	0.267	3.4		
300			0.086	1.8	0.163	2.4				
400			0.100	1.8	0.190	2.5				
500					0.213	2.5				

OPERATING PRESSURES

Steam is normally generated to pressures above atmospheric level. For space heating use, low-pressure generation goes up to 3 bar gauge; high-pressure generation starts at around 6 bar gauge. If, as is normally the case, the steam is generated primarily for other uses, much higher pressures may be present, in which case a pressure-reducing valve may be employed on the connection to the space heating system.

To ensure a measure of storage to cope with fluctuations in demand, steam is generated to a higher pressure than required, with pressure-reducing valves used locally at the offtakes to steam-operated plant and heat exchangers.

STEAM PIPE-SIZING PROCEDURES

Steam velocity in distribution mains ranges from 30 m/s for a dryness fraction above 0.8 to 60 m/s for superheated steam. This compares with up to 10 m/s for low-pressure air systems and 1.5–3.0 m/s for LTHW heating systems in black mild steel. As with LTHW and air distribution systems the sizing procedure may be done on velocity or pressure drop. A typical pressure drop for steam distribution (up to 65 mm nominal bore) is 225 Pa/m.

From an inspection of the steam pipe-sizing tables, the pressure loss per metre corresponding to a steam velocity of 30 m/s progressively reduces with increases in pipe size above 65 mm, such that at 250 mm nominal bore it is around 60 Pa/m. It is therefore important to check the steam velocity if pipe sizing is initially undertaken on pressure drop.

STEAM GENERATION

Considering a two-pipe pressurized system, steam is generated from water within the boiler under its own pressure with the crown and isolating valves closed until steam is present and the required operating pressure is reached. The isolating valve is then opened slowly and then the crown valve opened *slowly*, a quarter turn at a time, to avoid a sudden pressure loss within the generator. Furthermore, on start-up, the system is cold, and steam released condenses rapidly. Air and possibly water present in the system must be displaced by the condensing steam, and water and condensate must be allowed to travel at low velocity to avoid water hammer in the steam main, which can be a terrifying and expensive experience.

The warm-up time for a steam system is therefore extended. As steam supplies the system, water must be fed into the boiler to maintain its water level. As the boiler is pressurized the boiler feed water must also be pressurized to ensure entry. A boiler feed pump is used for this purpose. It will be a reciprocating pump or, more likely now, a multistage centrifugal pump.

The condensate is returned to the plant room under the residual pressure from the steam supply or via a pumping and receiving unit to the hotwell, which is open to atmosphere, and it is from this tank that the feed pump draws the make-up boiler water. Figure 5.1 shows a two-pipe steam system diagrammatically. The system shows three condensate returns to the hotwell, with one being diverted to waste due to contamination, which can occur in process work.

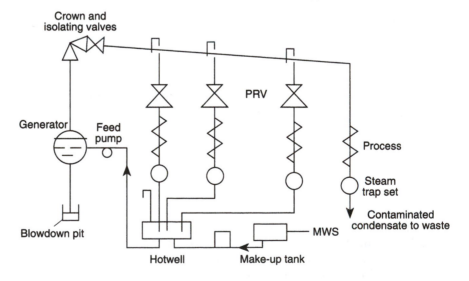

Figure 5.1   Two-pipe pressurized steam system.

Make-up water will be required in any event owing to evaporation at the hotwell. Water treatment will be required for the make-up water, and it is sometimes needed also in the boiler feed line.

Continuous or intermittent blowdown is required to rid the generator of suspended solids that are inevitably formed by the treatment processes from the impurities in the water.

STEAM PIPE SIZING ON VELOCITY

Typical velocities are given below for long straight pipe runs:

- superheated steam up to 60 m/s;
- saturated steam up to 40 m/s;
- wet steam with a dryness fraction above 0.8 up to 30 m/s.

Steam velocities for plant connections:

- steam connections to plant and equipment up to 15 m/s.

A formula for determining steam pipe diameter for a given flow rate and pressure can easily be derived from first principles:

volume flow rate = steam velocity × cross-sectional area of pipe

thus $\text{VFR} = u \times A = u(\pi d^2/4)$

$$\text{VFR} = Mv$$

Thus

$$Mv = u\left(\frac{\pi d^2}{4}\right)$$

from which

$$d = \sqrt{\frac{4}{\pi}} \times \sqrt{\frac{Mv}{u}}$$

and therefore

$$d = 1.1284 \times \sqrt{\frac{Mv}{u}} \quad \text{(m)} \tag{5.1}$$

As long as the dryness fraction q of the steam is above 0.8, specific volume $v = V_g \times q$. Otherwise specific volume v for the saturated vapour may be read directly from the steam tables as V_g.

Table 5.2 Some velocity factors for steam pipe sizing

Pipe diameter (mm)	Velocity factor K_6
15	5.733
20	3.059
25	1.935
32	1.079
40	0.786
50	0.595
65	0.284
90	0.152
175	0.037
200	0.030
225	0.023
250	0.018

Clearly this formula gives *actual* pipe diameter and not standard pipe diameter. The current *CIBSE Guide* adopts a velocity factor K_6 and a table for converting velocity factor values to *standard* pipe diameters.

From $Mv = u(\pi d^2/4)$:

$$u = \frac{4Mv}{\pi d^2} = Mv\left(\frac{4}{\pi d^2}\right)$$

If $K_6 = (4/\pi d^2)/1000$, then:

$$u = K_6 \times Mv \times 1000 \quad (\text{m/s}) \tag{5.2}$$

from which K_6 can be evaluated and standard pipe diameter obtained from the appropriate table in the *CIBSE Guide* or from Table 5.2.

Example 5.1

An HWS calorifier stores 600 litres of water at 65 °C from feed water at 10 °C with a regeneration of 1.5 h. If the primary connection is supplied with steam at 3 bar absolute 0.9 dry, determine the mass flow of steam required, the size of the steam connection and the actual steam velocity. Assume a maximum steam velocity of 10 m/s and that condensate leaves the primary heat exchanger at 130 °C. Ignore the inefficiency of the heat exchange.

Solution

The mass flow of steam M = (output in kW)/(heat given up by the steam in kJ/kg). Thus

$$M = \frac{Q}{dh} \quad \text{(kg/s)} \tag{5.3}$$

where

$$Q = \frac{600 \times 4.2(65 - 10)}{1.5 \times 3600} = 25.7\,\text{kW}$$

and

$$dh = h_{\text{w}} - h_{\text{f}130}$$

and from the steam tables

$$dh = (0.9 \times 2164 + 561) - 546 = 1963 \quad \text{(kJ/kg)}$$

Substituting using equation (5.3):

$$M = \frac{25.67}{1963} = 0.0131\,\text{kg/s}$$

So mass flow of steam = 0.0131 kg/s.
Now from equation (5.1):

$$d = 1.1284 \times \sqrt{\frac{Mv}{u}} \quad \text{(m)}$$

From the steam tables

$$V_{\text{g}} = 0.6057\,\text{m}^3/\text{kg}$$

Thus

$$v = V_{\text{g}} \times q = 0.6057 \times 0.9$$

and

$$v = 0.545\,\text{m}^3/\text{kg}$$

Substituting,

$$d = 1.1284 \times \sqrt{\frac{0.0131 \times 0.545}{10}}$$

from which

$$d = 0.03013\,\text{m}$$

The nearest standard pipe diameter in mild steel is 32 mm or 0.032 m.

By rearranging equation (5.1) in terms of velocity u and substituting $d = 0.032$ m, the *actual* steam velocity is evaluated to 8.88 m/s. Do you agree?

Summarizing:

- Mass flow of steam is 0.0131 kg/s.
- Standard pipe diameter in steel is 32 mm.
- Actual steam velocity is 8.88 m/s.

Further study
Using formula (5.2) and Table 5.2 you should now find the pipe diameter and actual velocity using the velocity factor K_6. Clearly the solutions will be similar.

Example 5.2
A steam main conveys 0.063 kg/s of saturated steam at an initial pressure of 2.7 bar absolute. The main is 70 m long and contains six welded bends and one stop valve. Assuming a maximum velocity of 40 m/s, determine:

(a)   the diameter of the main;
(b)   the final steam pressure;
(c)   the final steam velocity;
(d)   the initial steam velocity.

Data
Velocity pressure loss factors k: welded mild steel bends $k = 0.3$, stop valve $k = 6.0$.

Solution
Table 5.1 can be used for pipe sizing but the full tables in the *CIBSE Guide* are preferred.

(a)   Using equation (5.1), which determines actual pipe diameter, and substituting the data,

$$d = 1.1284 \times \sqrt{\frac{0.063 \times 0.6686}{40}} = 0.0366 \, \text{m}$$

The nearest standard diameter for steel pipe is 40 mm. Note the specific volume V_g of steam at 2.7 bar abs, from the steam tables, is 0.6686 m^3/kg.

This can be substituted as v directly into equation (5.1) as the steam is saturated.

(b)   You will now need access to the steam pipe sizing tables in the *CIBSE Guide* in order to proceed with the solution or refer to Table 5.1.

Given the mass flow is 0.063 kg/s and pipe diameter is 40 mm, $(Z_1 - Z_2)/L = 171$, and $l_e = 1.7$.

The total equivalent length

$$TEL = L + K_t \times l_e \quad (m)$$

The total velocity pressure loss factor is given by

$$k_t = \text{six welded bends} + \text{one stop valve}$$
$$= 6 \times 0.3 + 1 \times 6.0 = 7.8\,m$$

Thus

$$TEL = 70 + 1.7 \times 7.8 = 83.26\,m$$

Substituting in $dZ/L = 171$,

$$dZ = 83.26 \times 171 = 14\,237$$

From the rubric in the pipe-sizing tables, pressure factor $Z = P^{1.929}$ when P is in kPa. Thus

$$Z_1 = 270^{1.929} = 48\,989$$

Now

$$Z_1 - Z_2 = 14\,237$$

Substituting for Z_1,

$$48\,989 - Z_2 = 14\,237$$

from which

$$Z_2 = 48\,989 - 14\,237 = 34\,752$$

and

$$P_2 = {}^{1.929}\sqrt{34\,752} = 226\,kPa$$

Thus final steam pressure = 2.26 bar absolute.

The steam velocities can be obtained from either of the equations derived for actual pipe diameter d or velocity factor K_6.

(c)   From equation (5.1):

$$d^2 = 1.2733 \times \frac{Mv}{u}$$

From the steam tables, $V_g = 0.79\,\text{m}^3/\text{kg}$ at a final pressure of 2.26 bar absolute and pipe diameter $d = 0.04\,\text{m}$. Mass flow of steam is 0.063 kg/s and therefore by substitution,

$$\text{final steam velocity } u = 39\,\text{m/s}$$

(d)   From equation (5.2):

$$u = K_6 \times M \times v \times 1000$$

The velocity factor table from Table 5.2 gives $K_6 = 0.786$. Specific volume v at 2.7 bar absolute from the steam tables is $0.6686\,\text{m}^3/\text{kg}$ for saturated steam, and therefore by substitution:

$$\text{initial steam velocity } u = 33\,\text{m/s}$$

You should now confirm the solutions in parts (c) and (d).

Conclusion

You will notice that the final steam velocity is greater than the initial velocity. This is due to the effect of increasing specific volume with decreasing steam pressure. See if you get a similar final velocity by employing equation (5.2). The critical steam velocity in a section of steam distribution pipe is therefore the final velocity at the end of the pipe section.

The solution takes no account of heat loss from the steam main. It is of interest to know whether the quality of the steam suffers as it is transported over a distance of 70 m. If the pipe is efficiently lagged with 25 mm of insulation having a thermal conductivity of 0.055 W/mK, the heat emission from the main is 3146 W for a temperature difference of 107 K. This produces a dryness fraction at the end of the main of 0.997, which is close to saturated conditions at the final pressure of 2.26 bar absolute.

You might like to check this statement, in which case you will need the heat emission from the 40 mm pipe insulated as described, and this is 0.42 W/K per metre run.

Example 5.3

A steam ring main follows the perimeter of a factory at high level with connections coming off the ring as shown in Figure 5.2. Steam is supplied at 2 bar gauge 0.9 dry. Size the ring main on an initial velocity of 30 m/s and the branches on 10 m/s. Determine the steam velocity in the ring between branches 6 and 7 assuming all branches are in use. State any assumptions made.

Solution

When all the connections are in use the total mass flow rate of steam required in the ring main will be the sum of the individual steam flows and $M_T = 0.173$ kg/s. Adopting equation (5.1) and obtaining specific volume V_g from the steam tables at 3 bar absolute:

$$d = 1.1284 \times \sqrt{0.173 \times 0.6057 \times \frac{0.9}{30}} = 0.063 \, \text{m}$$

Thus the standard diameter for the ring main $d = 65$ mm.

The diameter for each of the branches may be determined in the same manner if the pressure loss along the ring is ignored. Determine these yourself and see if you agree with the standard pipe sizes listed in Table 5.3. What assumption has been made in determining the branch sizes? The mass flow required between branches 6 and 7 is 0.065 kg/s. Do you agree?

The ring main size is 65 mm. Rearranging equation (5.1) in terms of velocity:

$$u = (1.1284)^2 \times \frac{Mv}{d^2} = 1.2733 \times 0.065 \times 0.6057 \times \frac{0.9}{(0.065)^2} = 10.7 \, \text{m/s}$$

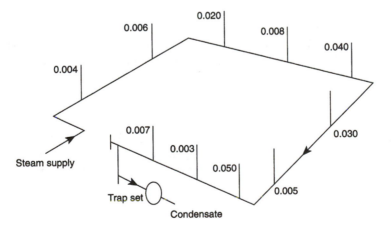

Figure 5.2   Ring main and mass flow rates.

Table 5.3 Example 5.3: standard pipe sizes

Branch no	Branch	Standard diameter (mm)	Actual velocity (m/s)
1	0.004	20	6.94
2	0.006	20	10.41
3	0.020	40	8.68
4	0.008	25	8.88
5	0.040	50	11.11
6	0.030	50	8.33
7	0.005	20	8.68
8	0.050	65	8.21
9	0.003	15	9.25
10	0.007	25	7.77

Steam velocity between branches 6 and 7 is 10.7 m/s. This is well below the initial velocity of 30 m/s because the flow rate has dropped from 0.173 kg/s just before branch 1 to 0.065 kg/s. The actual steam velocity just before branch 1 is 28.4 m/s. You should now confirm that this is so.

PIPE SIZING ON PRESSURE DROP

So far we have considered steam pipe sizing on velocity. However, in example 5.2, the final steam pressure was determined, and the pressure drop along the main can easily be calculated from $P_1 - P_2$: i.e. $270 - 226 = 44$ kPa. This can be expressed as a pressure drop per metre run, thus:

$$\frac{\mathrm{d}p}{\mathrm{TEL}} = \frac{44\,000}{83.26} = 528\,\mathrm{Pa/m}$$

Invariably steam is distributed slightly wet, and the recommended rate of pressure loss equivalent to approximately 30 m/s is 225 Pa/m for pipe sizes up to 65 mm. This figure will be used in the example that follows.

Example 5.4
The distribution steam main is shown in Figure 5.3. Steam is supplied at 500 kPa absolute, 0.9 dry. From the data, size the steam pipework and attempt to balance the branch with the index run in the choice of branch pipe size.

Data

Section	1	2	3
Mass flow	0.4	0.2	0.2
TEL	60	50	12

Figure 5.3   Steam distribution mains to two terminals.

Solution

Table 5.1 can be used for pipe sizing but the full tables in the *CIBSE Guide* are preferred. The index run for the steam distribution is that circuit with the greatest pressure drop and, in this case, includes sections 1 and 2 if the pressure drops at the terminals are similar. Initial pressure factor Z_1 is calculated from the initial pressure, which is 500 kPa, as 160 810. The approximate value of the final pressure factor Z_2 is obtained from $500 - (225/1000 \times (60 + 50))$ and equals 475 kPa. This is equivalent to a pressure factor of 145 661. The index pressure factor loss:

$$\frac{dZ}{L} = \frac{160\,810 - 145\,661}{60 + 50} = 138$$

Note the difference here from pipe sizing for water distribution: the pressure loss per metre is now pressure factor loss per metre. This takes account of the fact that steam is a compressible vapour whereas water is an incompressible liquid.

The index pipe diameters may now be selected from the pipe-sizing tables and the actual rates of pressure factor loss recorded. The initial and final pressure factor Z is calculated for each pipe section, so that Z_2 for section 1 becomes Z_1 for section 2. The given and calculated data are listed in Table 5.4.

Table 5.4   Example 5.4: given and calculated data

Section	1	2	3
M (kg/s)	0.4	0.2	0.2
TEL (m)	60	50	12
Available dZ/L	138	138	492
Diameter, d (mm)	90	65	50
Actual dZ/L	90	118	443
dZ	5400	5900	5316
Z_1	160 810	155 410	155 410
Z_2	155 410	149 510	150 094
P_1 (kPa)	500	491	491
P_2 (kPa)	491	481	482
v_2 (m³/kg)	0.3436	0.3507	0.3507
u_2 (m/s)	21	20	34.7

Having determined the actual final pressure in section 2 as 481 kPa (note the approximate value was estimated as 475 kPa), the final pressure in section 3 needs to be the same in order to balance the system. This enables the available dZ/L to be calculated as 492, and the table can then be completed.

Conclusion

Note that in section 1, the pipe size is above 65 mm but actual dZ/L is significantly lower than the available dZ/L based on 225 Pa/m.

The process of attempting to balance the system in the selection of pipe sizes is recommended for sizing steam pipe networks, as the use of regulating valves on a steam system is not an option, owing to the eroding effect of steam on valve seatings. The layout of the distribution pipework therefore assumes an important part of the design process for this reason. Ideally, for terminals having similar pressure drops, each needs to be the same distance from the steam supply point. From a practical viewpoint this is rarely achieved, and the balancing process is usually only partially successful. An important part of the sizing procedure is therefore to ensure that the pressure drop in the index run is kept low. This ensures that the first branch does not suffer a high pressure drop and hence an excessive steam velocity.

You will note that the final velocity in section 3 is excessive for wet steam, and this should initiate a review of the pipe layout to see whether branches 2 and 3 can be made more equidistant from the junction.

PRACTICAL PROBLEMS

A more practical approach to steam pipe sizing would be to consider an available steam supply pressure and condition and the needs of the terminal in terms of required steam pressure and condensate condition.

Example 5.5

A steam distribution main is shown in Figure 5.4 serving two terminals, each of which requires a steam pressure of 400 kPa via pressure-reducing valves (PRV) with condensate leaving as saturated water at 200 kPa. The initial steam condition is 500 kPa and saturated. Size the pipework assuming the lengths given to be total equivalent lengths and adopting a velocity of 30 m/s as the design parameter.

Determine also the pressure drop across each PRV. Assume also that there is negligible heat loss from the main and that the quoted pressures are absolute.

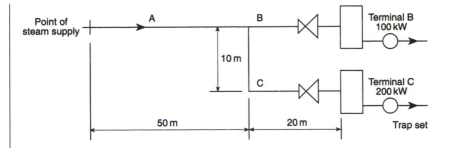

Figure 5.4 Example 5.5: diagrammatic layout of steam distribution.

Solution
Table 5.1 can be used for pipe sizing although the full tables in the *CIBSE Guide* are preferred. Because of the difference in pressure between the steam supply point and the point of use $(500 - 400 = 100\,\text{kPa})$, there is a measure of storage available in the distribution main in the event of fluctuations in load requirements.

As there is no heat loss to be accounted for along the steam main, heat content is constant along its length at $500\,\text{kPa}$, and $h_\text{g} = 2749\,\text{kJ/kg}$. The heat in the saturated condensate leaving the terminals at $200\,\text{kPa}$, $h_\text{f} = 505\,\text{kJ/kg}$. From these data the mass flows can be calculated:

$$M_\text{B} = \frac{100}{2749 - 505} = 0.0446\,\text{kg/s}$$

$$M_\text{C} = \frac{200}{2749 - 505} = 0.089\,\text{kg/s}$$

and therefore $M_\text{A} = 0.134\,\text{kg/s}$. From equation (5.1):

$$d = 1.1284 \times \sqrt{\frac{0.134 \times 0.3748}{30}} = 0.046\,\text{m}$$

The nearest standard size for pipe A, $d = 50\,\text{mm}$.

The pressure P_2 at the first branch can now be determined. Knowing the mass flow and pipe size in A, from the pipe-sizing tables: $dZ/L = 211$, $dZ = 211 \times 50 = 10\,550$ and $Z_2 = Z_1 - dZ$. Thus $Z_2 = 161\,000 - 10\,550 = 150\,450$, from which $P_2 = 483\,\text{kPa}$.

Pipe B may now be sized, and

$$d = 1.1284 \times \sqrt{\frac{0.0446 \times 0.388}{30}} = 0.027\,\text{m}$$

The nearest standard diameter for pipe B, $d = 32\,\text{mm}$.

Pressure at the PRV B can now be determined. Knowing the mass flow and pipe size in B, from the pipe-sizing tables, $dZ/L = 200$, $dZ = 200 \times 20 = 4000$, and $Z_2 = Z_1 - dZ$. Thus $Z_2 = 150\,450 - 4000 = 146\,450$, from which $P_2 = 476\,kPa$. Therefore the pressure drop across the PRV on pipe B= $476 - 400 = 76\,kPa$.

Determination of the size for pipe C:

$$d = 1.1284 \times \sqrt{\frac{0.089 \times 0.388}{30}} = 0.383\,m$$

The nearest standard diameter for pipe C, $d = 40\,mm$.

Determining the pressure at PRV C: knowing the mass flow and pipe size, from the pipe-sizing tables, $dZ/L = 321$, $dZ = 321 \times (10 + 20) = 9630$, $Z_2 = 150\,450 - 9630 = 140\,820$, from which $P_2 = 467\,kPa$. Therefore the pressure drop across the PRV on pipe C $= 467 - 400 = 67\,kPa$.

Conclusion

It is important to note that the steam pressure available to take the condensate back to the hotwell is 200 kPa absolute, and therefore the effective steam pressure available is 100 kPa gauge. The actual final velocities in each of the pipe sections can now be evaluated:

$$u_A = 26.5\,m/s$$
$$u_B = 21.8\,m/s$$
$$u_C = 28.5\,m/s$$

You should now confirm these actual final velocities.

USE OF THE MODULATING VALVE ON THE STEAM SUPPLY

A difficulty encountered in returning the condensate using the available steam pressure occurs when a modulating valve is employed on the steam main serving the terminal. As the valve modulates, the orifice reduces in size, although the steam velocity tends to increase, thus counteracting the attempt to reduce the steam flow. For satisfactory operation, the downstream pressure must therefore not be less than 0.6 of the steam pressure upstream of the valve.

When the valve has modulated to its maximum partially closed position the steam pressure available downstream to drive the condensate back to the hotwell may not be sufficient, in which case the terminal being served becomes waterlogged. In these circumstances, a mechanical condense pump operated by the steam on the live side of the modulating valve may be required. Alternatively a mechanically or electrically driven condensate receiver and pumping unit may be the solution. You should make yourself familiar with these items of plant.

Example 5.6
The diagram in Figure 5.5 shows the principal features of a steam supply serving a heat exchanger. Size the steam main on a velocity of 20 m/s and the condensate return on one-tenth of the available pressure. Determine the discharge pressure of the condensate at the entry to the hotwell.

   Consider the effect that a modulating valve located in the steam supply would have on the return of the condensate under conditions of maximum modulation. All pressures quoted are absolute with the steam trap handling condensate at 127.4 °C.

Solution
With the modulating valve fully open, the mass flow rate M can be determined using data from the steam tables and assuming that the condensate in the steam trap is close to saturation. Thus

$$M = \frac{80}{546 + 0.9 \times 2174 - 535} = 0.041 \text{ kg/s}$$

$$d = 1.1284 \times \sqrt{\frac{0.041 \times 0.6686 \times 0.9}{20}} = 0.04 \text{ m}$$

Steam supply pipe diameter $= 40$ mm. The available pressure for sizing the high-level condensate pipe, P_c, will be $P_b -$ static lift. From the steam tables the specific volume v_f of water at 127.4 °C, which is the saturation-temperature at 250 kPa, is 0.001066 m³/kg. The density is therefore 938 kg/m³. Thus

$$P_c = 250 - \frac{3.22 \times 938 \times 9.81}{1000} = 220 \text{ kPa abs}$$

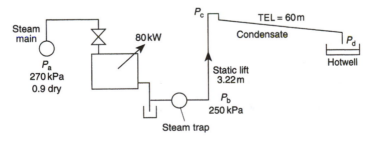

Figure 5.5 Example 5.6: principal features of a steam supply and condensate return to the hotwell in the plant room.

and saturation temperature $= 123.3\,°C$, so the water at this point will partially evaporate into flash steam. Gauge pressure at P_c is therefore 120 kPa, and if all the pressure is absorbed in sizing the condense return the pressure loss per metre will be

$$\frac{120\,000}{60 \times 10} = 200\,Pa/m$$

Note that the figure of 1/10 of the available pressure drop given in the data to this example allows for the fact that the condensate return will have to handle condensate, flash steam and air in varying unknown proportions.

Clearly, it is not possible to absorb all the availiable 120 kPa, as there must be sufficient residual pressure P_d in the condensate outlet at the hotwell for discharge into the tank. From the *CIBSE* pipe-sizing tables for X-grade copper, a mass flow of 0.041 kg/s at an available pressure drop of 200 Pa/m, a 22 mm pipe has an actual pressure loss of 14 Pa/m. This will clearly leave sufficient residual pressure for discharge into the tank.

If the steam valve modulates down to its minimum downstream pressure of 0.6 of 170 kPa gauge, it will have the effect of reducing P_c to 52 kPa gauge and the available pressure drop per metre along the condensate return will be reduced to

$$\frac{52\,000}{60 \times 10} = 87\,Pa/m$$

However modulation of the steam valve will clearly reduce the flow rate below 0.041 kg/s so this should not alter the size of the condense pipe or the return of the condensate to the hotwell.

Conclusion

In this case, the modulation of the steam control valve should not affect the return of the condensate, but you can see the potential effect that it could have at a lower steam supply pressure with insufficient pressure left for the return of the condensate to the hotwell. It is in such cases that a mechanical pump referred to earlier would be required.

Table 5.5 recommends a condensate pipe size of 25 mm for a steam pipe size of 40 mm; you can see here a discrepancy in sizing the condense main. Remember this results from the generally unknown quantities of air and flash steam that may be present in the condense pipe, particularly during system start-up from cold. For this reason, it may be advisable to err towards the larger size of pipe.

Table 5.5   Recommended condensate pipe sizes

Steam d (mm)	20	25	32	40	50	65	80	100	150
Recommended condensate d	15	20	20	25	32	32	50	65	80

STEAM TRAPS

Steam traps are required to inhibit the flow of steam as it passes into the heat exchanger so that it gives up its latent heat. They are also required to prevent live steam entering the condense main.

In most space heating applications, the steam trap must evacuate the condensate as it forms in the heat exchanger. There are a number of different types of steam trap available: float operated, inverted bucket, thermostatic, liquid expansion, bimetallic and thermodynamic. The capacity of the steam trap depends upon the size of its orifice, condensate temperature and the pressure drop across the trap.

As condensate temperature rises, density falls and flow rate is increased. As the pressure drop across the trap increases so does the flow rate. If the condense pipe rises after the trap, back pressure on it is increased, and the pressure drop across the steam trap is reduced, thus reducing the rate of flow.

You should familiarize yourself with the various types of steam trap on the market, and identify which to use for different heating apparatus. [Spirax Sarco Ltd., Cheltenham.]

Each heater must be fitted with its own steam trap. Groups of similar heaters with similar traps can be connected to a common condensate return. Dissimilar steam equipment and associated traps must not be connected to a common condensate return. This is due to the way traps operate. Some are intermittent in operation; others are continuous. This and variations in steam pressure required by different apparatus can cause waterlogging in the heat exchangers operating at a lower pressure when different apparatus are interconnected.

This identifies one of the disadvantages of steam systems, in that a number of condensate returns from different apparatus must be run separately back to the hotwell. An alternative is to employ a receiver and pumping unit into which they are discharged, allowing a single condensate return from the unit to the hotwell.

USE OF THE VACUUM BREAKER

This is considered, for example, on a steam-fed air heater battery, where the duct thermostat is controlling a modulating valve on the steam supply.

As the valve modulates in response to the duct thermostat, the steam in the battery condenses, and this can cause a partial vacuum, which results in the condensate not being discharged at the steam trap. The heater battery partially fills with the condensate, thus causing a severe temperature gradient in the air stream. The vacuum breaker prevents this occurrence. However, if it draws in very low-temperature outdoor air to counter the partial vacuum it can cause freezing conditions within the battery. It is therefore important to ensure that the vacuum breaker is located to avoid this possibility. [Spirax Sarco Ltd., Cheltenham.]

RELAY POINTS

Particularly on system start-up and to a much smaller extent during system operation, condensate will form in the steam pipework. This must be removed at regular intervals to keep the steam mains largely free of condensate. There are two kinds of relay point, which are shown in Figure 5.6.

Figure 5.6(a) shows relay points for use on a distribution main. Note that the condensate is aided by steam flow and gravity towards the collection point, with the steam main maintaining the same level. Figure 5.6(b) shows a relay point for use in a plant room.

PRESSURE-REDUCING VALVES

Various apparatus using steam require specific pressures to operate at optimum outputs. It is therefore normal to employ pressure-reducing valves (PRV) on the live steam supply locally to steam-heated equipment, where it can be reduced to the specified pressure. The data required to size a PRV include the mass flow, expected steam quality, upstream and downstream steam pressures. Steam flow through the PRV follows an adiabatic process, during which the quality of the vapour tends to improve. Consider the following example.

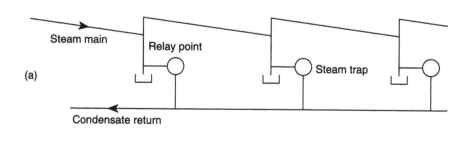

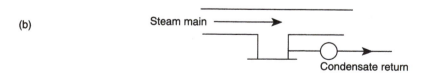

Figure 5.6 Two types of relay point.

Example 5.7
Steam at 10 bar absolute 0.93 dry is fed through a PRV, which reduces its pressure to 2.0 bar absolute. Determine the condition of the steam on the downstream side of the valve and the temperature drop sustained across the valve.

Solution
The heat content of the steam on the live side of the PRV, $h_w = 763 + 0.93 \times 2015 = 2637\,\text{kJ/kg}$. At 2.0 bar the heat content of saturated steam, $h_g = 2707\,\text{kJ/kg}$. As the process is adiabatic, the heat content of the steam on the low-pressure side is the same as that of the live steam, namely 2637 kJ/kg. Clearly then the steam still has a dryness fraction q, and at the low pressure $h_w = h_f + q \times h_{fg}$. Thus

$$2637 = 505 + q \times 2202$$

from which

$$q = 0.97$$

Summary
The quality of the steam has improved from a dryness fraction of 0.93 to 0.97. From the steam tables, the saturation temperatures at 10 bar and 2.0 bar are 180 °C and 120 °C respectively. The temperature drop across the PRV is therefore 60 K. The heat content of the steam on either side of the PRV is 2637 kJ/kg.

SIZING THE CONDENSATE RETURN

On system start-up from cold, the condensate mains will have to handle air and water. If the condense is lifted on the downstream side of the steam trap, there is a good possibility that some of the condensate at the top of the rise will flash back into steam during normal operation.

The condensate pipe therefore has to handle a mixture of air, water and flash steam in varying amounts. It is therefore almost impossible to size the condense pipe accurately. It has been found that if it is sized one to three sizes below that of the steam main that it accompanies, it performs its tasks satisfactorily. Table 5.5 is a guide.

Example 5.6 identifies an alternative approach where the pressure loss sustained along the condense main is calculated as 10 times that for LTHW systems. Thus $dp = 10(\text{pd per metre} \times \text{TEL})$, from which available pd per metre $= dp/(10 \times \text{TEL})$.

A further alternative recommends that the condense main is sized on three times the normal hot water discharge. Both these alternatives tend to yield smaller pipe sizes than those in Table 5.5.

Example 5.8
Figure 5.7 shows a branch taken from a steam main to serve a heat exchanger via a PRV. Pressures quoted are in absolute units.

- Size the branch off the high pressure main on 225 Pa/m.
- Size the low pressure connection on 9 m/s.
- Size the PRV.
- Size the condensate return.

Solution
The mass flow required at the heater can be calculated from the given data and steam tables and $M = 0.061$ kg/s. Adopting the velocity factor formula, $K_6 = 0.313$, and therefore from the velocity factor (Table 5.2) $d = 65$ mm. The size of the LP steam connection is 65 mm.

The listed value for K_6 in the velocity factor table is 0.284, from which the actual velocity of the steam in the heater connecting pipe can be calculated. You should now calculate the actual steam velocity and compare it with the adopted velocity of 9 m/s.

dZ/L for the HP branch is calculated to be 189 for a flow rate of 0.061 kg/s, and from Table 5.1 $d = 40$ mm for a dZ/L of 162. The size of the HP branch is 40 mm. This allows the determination of pressure factor Z_2, from which P_2 at the PRV inlet is determined as 686 kPa.

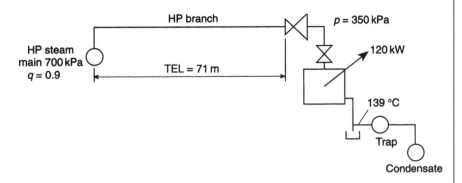

Figure 5.7   Example 5.8: steam connection to heater.

The data for sizing the PRV are therefore: inlet pressure = 686 kPa, outlet pressure = 350 kPa, mass flow = 0.061 kg/s. From Table 5.5, the estimated size of the condensate return is 32 mm.

Do you agree with these solutions?

You should now verify these solutions following the procedures in the solutions to earlier examples.

5.3   Steam generation and distribution

There are two ways by which steam may be distributed:

(1)   *Distribution by high pressure from the generator with PRVs located in the branches serving the appliances*: This allows apparatus operating at different pressures to be served from one generating plant. Operating the generator at a pressure in excess of the maximum required offers a measure of storage in the distribution mains, which will be smaller in size compared with low-pressure steam distribution.

(2)   *Distribution by low pressure with the necessary allowance for pressure loss along the index run*: This may allow the use of lower-specification plant with correspondingly lower capital cost. However, it is likely to cope only with apparatus having a common operating pressure, and there is no inherent margin of steam in the distribution mains to offset a sudden increase in demand. Furthermore, the condensate return may require assistance from pumping and receiving units, as there may be insufficient steam pressure left at the index terminal to drive the condense back to the hotwell.

Case Study 5.1

A feasibility study is required for steam distribution on a sloping site to three buildings. Each building requires 1 MW of LTHW heating and 10 000 litres/h of HWS. The minimum pressure in the heating and HWS systems is 1.0 bar gauge.

The study shall include the following outcomes:

(a)   generator and distribution pressure;
(b)   steam flow rate to each building;
(c)   boiler power and equivalent evaporation from and at 100 °C;
(d)   recommended method for returning the condensate;
(e)   estimates of steam distribution and condensate return sizes;
(f)   estimation of the linear expansion in the distribution main and recommendations on methods of accounting for it.

The site is shown in Figure 5.8.

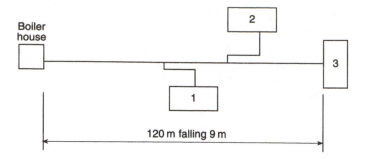

Figure 5.8   Case study on steam distribution, buildings 1, 2, 3. Plan view.

SOLUTION

Clearly there is more than one solution to this case study, if only in the choice of steam pressure for the distribution mains. The distribution steam pressure selected here will be the minimum for satisfactory system operation. It is an easy matter to raise the steam pressure above the minimum to provide for a measure of storage.

Maximum steam pressure at each building

If the temperature controls fail, the secondary water for both the heating and HWS should not boil. Thus the maximum steam temperature should be 10 K below boiling point at the minimum water pressure of 1.0 bar gauge. From the steam tables saturation temperature at 2.0 bar absolute is 120 °C, and allowing a 10 K antiflash margin, leaves a maximum steam temperature of 110 °C. This corresponds to a steam pressure of 1.4 bar absolute at each building. It is apparent that the steam supply to each building requires a PRV, as 1.4 bar absolute (0.4 bar gauge) is of little use as a distribution steam pressure from the boiler house.

The temperature drop between primary and secondary media should be above 20 K to ensure positive heat exchange. If the LTHW heating mean water temperature is 80 °C, $dt = 110 - 80 = 30$ K. Similarly if HWS storage temperature is 65 °C, $dt = 110 - 65 = 45$ K. Both temperature drops are satisfactory.

If the condensate is to be returned under its own pressure, steam distribution pressure will depend upon the minimum pressure required to return the condensate to the boiler house from the index terminal. The minimum initial pressure required in the condense return therefore, adopting a pd of 300 Pa/m (remember the pipe is sized on 30 Pa/m if the factor of 10 is used) and making an allowance of 30% for fittings on straight pipe is given by the following:

Initial pressure $P = \text{TEL} \times \text{pd/m} = 120 \times 1.3 \times 300 = 46\,800\,\text{Pa}$

Taking density as $940\,\text{kg/m}^3$

$$\text{the initial head } h = \frac{46\,800}{940 \times 9.81} = 5\,\text{m}$$

In view of the sloping site:

$$h = 5 + 9 = 14\,\text{m}$$

This corresponds to a minimum initial pressure in the condensate return of

$$P = 14 \times 940 \times 9.81 = 130\,\text{kPa gauge}$$

To ensure satisfactory operation, and allowing for the pressure drop across the steam heat exchanger in the index terminal, minimum final steam pressure available at the PRV of 1.5 bar gauge, 2.5 bar absolute is therefore required on the live side of the PRV. Thus the pd across the index PRV will be from 2.5 to 1.4 bar absolute. Steam from the live side of the PRV will now have sufficient pressure to drive the condensate back to the hotwell via a pumping and receiving unit.

Allowing for a pressure drop of 225 Pa/m along the steam distribution main, minimum boiler pressure can now be established.

(a) Minimum generator pressure $= ((225 \times 120 \times 1.3)/(1000)) + 250 = 285\,\text{kPa}$ abs. The demand at each building is as follows: heating 1.0 MW and HWS 10 000 litres/h. It can be assumed that the quality of the steam on the downstream side of the PRV is 0.95, and from the steam tables at 1.4 bar absolute:

$$h_w = 458 + 0.95 \times 2232 = 2578\,\text{kJ/kg}$$

If condensate leaves at 107 °C:

$$h_f = 449\,\text{kJ/kg}$$

Thus the mass flow for the heating will be

$$M = \frac{1\,\text{MW} \times 1000}{2578 - 449} = 0.47\,\text{kg/s}$$

The output for the HWS:

$$Q = \frac{10\,000}{3600} \times 4.2 \times (65 - 10) = 642\,\text{kW}$$

Thus the mass flow for the HWS = 642/(2578 − 449) = 0.3 kg/s. Total net mass flow = 0.47 + 0.3 = 0.77 kg/s. Allowing for inefficiency of heat transfer of 20%, total mass flow to the index building M = 0.92 kg/s. The steam serving the other two buildings on the downstream side of the PRVs is likewise 1.4 bar absolute.

(b)   Steam flow rate to each building is 0.92 kg/s. Total mass flow from the generator will be 3 × 0.92 = 2.76 kg/s. If the boiler generates steam at 0.95 dry from feed water at 80 °C, the boiler power required will be

$$M \times dh = 2.76 \times (556 + (0.95 \times 2168) - 335)$$

$$= 2.76 \times 2281 = 6295\,kW$$

(c)   Boiler output power required is 6300 kW at an operating pressure of 185 kPa gauge. Equivalent evaporation from and at 100 °C is used as the basis for comparing different steam generators for the purposes of selection. It is equivalent to ((water evaporated in kg/s) × dh/(fuel input in kg/s))/2257 at atmospheric pressure, where the latent heat of vapourization is 2257 kJ/kg. Assuming boiler efficiency of 75% and a gross calorific value for the fuel of 40 MJ/kg,

$$\text{Fuel input} = \frac{6295}{40\,000 \times 0.75} = 0.21\ \text{kg/s}$$

$$\text{water evaporated} = 2.76\ \text{kg/s} \quad \text{and} \quad dh = 2281\ \text{kJ/kg}$$

Thus equivalent evaporation from and at 100°C = ((2.76 × 2281)/0.21)/2257 = 13.28 kg water/kg fuel.

The limitation of this comparison is that the fuel must be common to each generator being considered. The condensate return will need to be via a receiver and a mechanical pumping unit operated from the steam supply on the live side of the PRV at a pressure of 150 kPa gauge. This was established in determining the minimum generator pressure. Alternatively the pumping unit can be an electrically driven pump in which the net pressure developed would need to be equivalent to a lift of 14 m (130 kPa) and the flow rate equivalent to the condensate flow of 0.92 kg/s.

(d)   Methods for returning the condensate are as follows: The size of the steam distribution main will be based upon the initial and final pressures, where

$$Z_1 = P_1^{1.929} = 285^{1.929} = 54\,375$$

$$Z_2 = P_1^{1.929} = 250^{1.929} = 42\,231$$

$$\frac{dZ}{L} = \frac{12\,144}{120 \times 1.3} = 78$$

Initially this dZ/L will be used to size each section of the steam main. From the steam pipe-sizing tables for a mass flow of 2.76 kg/s, the pipe size is 200 mm, for which $dZ/L = 61$. As the pipe size is in excess of 65 mm, steam velocity must be checked, and from equation (5.1):

$$u = (1.1284)^2 \times \frac{Mv}{d^2} = 1.2733 \times \frac{2.76 \times 0.6358 \times 0.95}{(0.2)^2} = 53\,\text{m/s}$$

Maximum velocity for wet steam is 30 m/s. The reason for the excessive velocity is the use of 225 Pa/m in determining the minimum boiler pressure. With such large-diameter pipe a figure of 60 Pa/m would be more appropriate.

To be sure of not exceeding the maximum velocity, the distribution mains will be sized on 30 m/s. Thus for a mass flow of 2.76 kg/s, a specific volume V_g of 0.6358, a dryness fraction of 0.95, and adopting equation (5.1), $d = 266$ mm. For a mass flow of 1.84 kg/s, a specific volume of 0.6772 and a dryness fraction of 0.95, $d = 224$ mm. For a mass flow of 0.92 kg/s, a specific volume of 0.7186 and a dryness fraction of 0.95, $d = 163$ mm. The specific volumes were taken at the beginning, mid-point and end of the distribution main respectively.

(e) The suggested standard sizes of the steam distribution main are therefore 250 mm, 225 mm and 175 mm. It is left to you to check the actual steam velocities in each section. The size of the condense main from the index terminal is based on a mass flow of 0.92 kg/s and a pd of one-tenth of the adopted value taken as 300 Pa/m. Thus at a maximum 30 Pa/m and using copper pipe to table X, $d = 67$ mm, for which pd = 14 Pa/m.

Three separate condensate returns, each at 67 mm diameter: You will find that the size of the condensate return is increased if the alternative estimate is adopted by extending Table 5.5 pro rata. What would you recommend?

There are in fact two options here for returning the condensate. One is to return each of the three condensate pipes back to the hotwell separately. The other is to return the condensate from building 3 into the receiver and pumping unit of building 2, and the common condensate from buildings 3 and 2 to the receiver and pumping unit of building 1, from which the condensate common to the three buildings is returned to the hotwell. This method will save on the length of copper used but at the expense of increases in pipe size to 76 mm from building 2 to 1 and to 108 mm from building 1 back to the boiler house. Do you agree with these condense pipe sizes if they are based on a maximum of 30 Pa/m: i.e. one-tenth of 300 Pa/m adopted.

Figure 5.9 shows the arrangement around the receiver and pumping unit in building 1. A flash steam exhaust pipe fitted with an exhaust head is taken from the receiver to outdoors to stabilize the pressure in

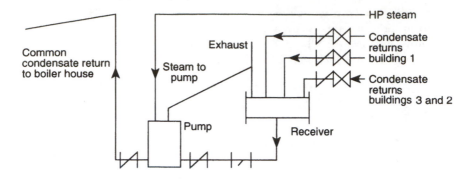

Figure 5.9   Receiver and mechanical pumping unit in building 1.

the receiver. This may help you to decide on the sizes of the condensate returns.

(f)   *Linear pipe expansion (for detailed consideration refer to Chapter 13)*: The total expansion from cold along the steam distribution main is obtained from expansion coefficient × length × temperature rise.

The coefficient for mild steel is 0.000012 m/mK, and the amount of linear expansion = 0.000012 × 120(132 − 10). Therefore the expansion = 0.176 m or 176 mm. This must be accounted for either through changes in the direction of the pipe in the form of expansion loops or double sets, or by employing axial compensators. In either case, the pipe will require anchoring on either side of the expansion device and at the entry to each building. The pipes will also require adequate pipe guides to control the expansion along the pipe axis and, where necessary, to control the lateral movement of the pipe. Vertical pipe movement must be avoided to ensure against condensate collecting at low points that are not fitted with a steam trap and high points where air or gases will collect.

The condense return is in copper, which has a coefficient of linear expansion of 0.000018 m/mK. Over the full length the amount of expansion will be

$$\text{expansion} = 0.000018 \times 120(109 - 10) = 0.214\,\text{m}$$

which is 214 mm.

Conclusion of the solution to the case study

Hopefully you have followed the rather tortuous routes to the solutions. You may have noticed that the minimum generator pressure could now be reduced slightly, as the pressure drop along the steam main was taken as 225 Pa/m when subsequently, owing to the large size of the pipe, a pressure of 60 Pa/m would have been appropriate. However, it is not considered necessary to re-work the solutions for such a small drop in boiler pressure.

The final topic for consideration is flash steam recovery and use. The use of a flash steam recovery system will improve the overall efficiency of the steam raising plant.

FLASH STEAM RECOVERY

High-pressure condensate coming off apparatus requiring HP steam has substantial heat content, which can be used effectively before returning it to the hotwell. Effective use improves the overall efficiency of the system.

The condensate can be used in a heat exchanger or a flash steam recovery vessel. Figure 5.10 shows a flash steam recovery vessel with its appropriate connections. You will notice that there is a high-pressure make-up line interconnected with the low-pressure flash steam. This ensures that the low-pressure demand is always met. The system is self-regulating in operation. You should now confirm this statement by studying Figure 5.10.

To obtain maximum useful heat from the HP condensate, the pressure of the flash steam must be as low as possible, which means that it should preferably be used locally to the flash vessel.

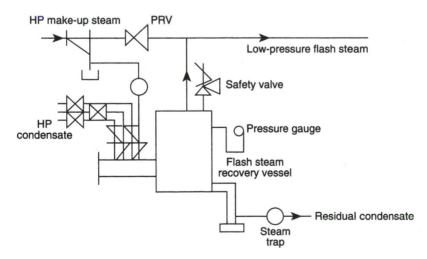

Figure 5.10 The flash steam recovery vessel and connections.

Example 5.9

0.22 kg/s of HP condensate at 195 °C is put through a flash steam recovery vessel, where it expands to a pressure of 0.5 bar gauge 0.9 dry.

Determine the mass flow of flash steam derived from the HP condensate, the potential output of this flash steam, the mass flow of residual condensate, and the mass flow of high-pressure steam make-up if the low-pressure demand rises to 100 kW. Assume that the HP steam is saturated.

Solution

Mass flow of flash steam:

$$M_f = \text{mass flow of HP condensate} \times \frac{\text{difference in } h_f}{\text{latent heat at LP}}$$

$$M_f = M\left(\frac{dh_f}{qh_{fg} \text{ at LP}}\right)$$

and from the steam tables

$$M_f = 0.22\left(\frac{830 - 467}{0.9 \times 2226}\right) = 0.04 \text{ kg/s}$$

Potential output of flash steam

$$Q_f = M_f \times q \times h_{fg} = 0.04 \times 0.9 \times 2226 = 80 \text{ kW}$$

$$\text{Residual condensate} = 0.22 - 0.04 = 0.18 \text{ kg/s at } 111.4 \text{ °C}$$

$$\text{High-pressure make-up mass flow} = \frac{\text{increase in demand}}{\text{heat given up}}$$

$$\text{Increase in demand} = 100 - 80 = 20 \text{ kW}$$

Heat given up = heat in HP steam – heat in LP condensate, and from the steam tables:

$$\text{Heat given up} = 2790 - 467 = 2323 \text{ kJ/kg}$$

Thus high-pressure make-up = 20/2323 = 0.00861 kg/s.

5.4   Chapter closure

This completes the work on steam heating systems. You should now attempt the steam pipe-sizing solution in Chapter 1. You are now able to undertake initial design studies on the use of steam as a means of space heating for a given site. Conversion from steam to LTHW is done using non-storage (heating) and storage (HWS) calorifiers, and the principle is covered in Chapter 4. You also have the skills now to size systems and plant, and to check limiting steam velocities as well as the pressures needed for the return of the condensate to the hotwell in the plant room. If steam is generated to a high pressure, perhaps for uses other than space heating, you are now able to consider the reuse of the high-temperature condensate in the form of flash steam.

6 Plant connections and controls

Nomenclature

BEMS	building energy management system
CPU	central processing unit
CTCV	constant temperature constant volume
CTVV	constant temperature variable volume
CVVT	constant volume variable temperature
CWS	cold water service
DDC	direct digital control
F&E	feed and expansion
HWS	hot water service
LAN	local area network
MWS	mains water service
OEM	original equipment manufacturer
TRV	thermostatic radiator valve
VDU	visual display unit
M	meter (electricity, fuel, heat)

6.1 Introduction

This chapter focuses upon the way in which systems are put together and automatically controlled. At an early stage in the design process for space heating and water services, it is necessary to identify how the building should be zoned, both in terms of time scheduling (those areas whose operating periods may differ) and in terms of temperature control (dependent upon building orientation, exposure and uses to which different parts of the building are going to be put). Figures 6.1 and 6.2 illustrate *zoning diagrams*.

Vertical zoning may be required, for example, to respond to the effects of building orientation. *Horizontal zoning* may be required, for example, to offset the effects of exposure of a multi-storey building where the upper storeys are subject to a more severe climate.

Zoning for time scheduling and temperature control marks a major step in the design process, for it implies knowledge on the part of the design team of the client's needs and the architect's vision of the form that the building will take.

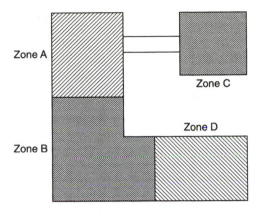

Figure 6.1   Site plan of building showing time-scheduled zones.

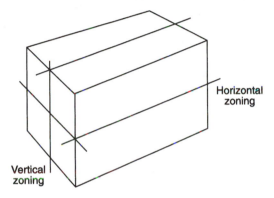

Figure 6.2   Isometric view of building showing zoning for temperature control.

Following the completion of the zoning plans, the next stage in the design process is to work up a schematic diagram that identifies the services and plant for the project. This chapter focuses on this issue. It will introduce the subject in parts, which can then easily be selected and assembled to form the schematic diagram for the project in hand.

6.2   Identifying services and plant

BOILER PLANT HEADER CONNECTIONS

Single or multiple boilers should always be connected to either a mixing header or separate flow and return headers. This allows heating circuits to be independently connected to the boiler plant, and facilitates the maintenance and breakdown of individual circuits and boilers. It is for this reason an indicator

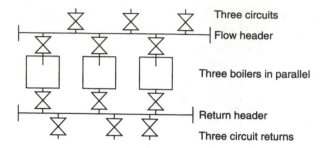

Figure 6.3   Boilers connected in parallel, serving three circuits.

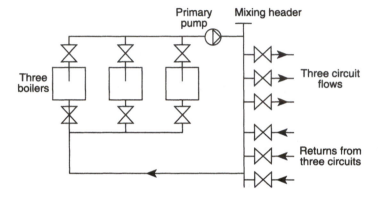

Figure 6.4   Three boilers connected to a mixing header, serving three circuits.

of good engineering practice. Figure 6.3 shows plant connected to flow and return headers.

The use of the mixing header has been reinstated since the introduction of modular boilers, and it is normally applied to gas-fired appliances. Figure 6.4 shows a typical layout.

Modular boilers are usually connected as shown in Figure 6.4. Note the use of a primary pump to ensure circulation through the boiler circuit. 'Secondary' pumps are required on each of the three heating circuits. Sequenced boiler control is employed to provide maximum efficiency of plant use.

BOILER PLANT OPEN VENT AND COLD FEED CONNECTIONS

When multiple boilers are used it is not necessary to connect an independent open vent or feed and expansion pipe to each boiler. This clearly saves cost and space, and Figure 6.5 shows typical connections to multiple boiler plant.

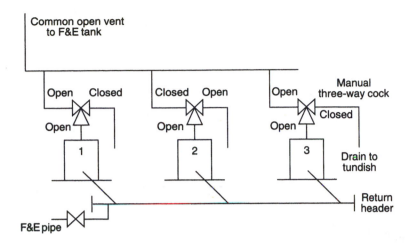

Figure 6.5 Open vent and cold feed connections to multiple boilers. Boiler return isolating valves omitted.

Normal position for the manual three-way cock is indicated for boilers 1 and 3. If, for example, boiler 2 needs to be isolated from the common open vent, the three-way cock is positioned as shown, with the boiler port and drain port open and the open vent port closed. This allows release of the static pressure within the boiler, and if draining down is necessary, air can enter the drain port, thus facilitating draining down. The remaining boilers are unaffected. The feed and expansion pipe is connected into the return header via a lock shield isolating valve.

BOILER RETURN TEMPERATURE PROTECTION

Heating systems with CVVT control are particularly subject to low return temperatures in mild weather. This can have a detrimental effect on the boiler by inducing condensation in the flue gas in the area of the return connection. Low return temperatures without corresponding modification to boiler flow temperature can also cause differential expansion within some boilers and consequent stress in the boiler metal. The boiler manufacturer will advise.

Figure 6.6 shows means of protection. When the thermostat senses low-temperature water in the boiler return, it energizes the pump and cuts it out on a predetermined higher temperature.

Figure 6.7a and b shows alternative arrangements when there is a constant volume flow through the boiler plant and there is still a requirement for protection. The regulating valve on the bypass pipe will need careful balancing, as it is subject to the full system pump pressure. In both cases the

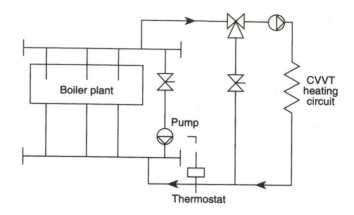

Figure 6.6   Boiler low return temperature protection.

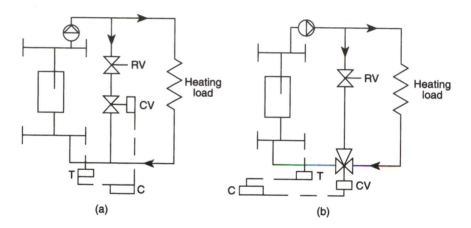

Figure 6.7   Boiler return protection for constant flow through the boiler plant: RV, regulating valve; CV, control valve; T, thermostat; C, controller.

control valves respond to the immersion thermostats when low-temperature water is detected, by opening to ensure flow through the bypass. On rise in temperature the controller closes the valve to bypass.

CONDENSING BOILERS

These boilers are specifically designed to operate on low return water temperature so that condensation can form on the flue gas side of a second heat exchanger. In the process, further sensible heat and latent heat of condensation from the flue gas are given up, thus increasing the thermal efficiency of the boiler. The secondary heat exchanger can be integral or separated

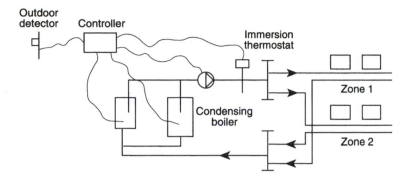

Figure 6.8 Application of the condensing boiler with direct weather compensation giving CVVT control.

from the primary heat exchanger in the boiler. If it is separated the boiler is called a *split condensing unit*, in which the primary and secondary heat exchangers can operate at different temperatures and flow rates. This can be put to advantage by allowing the primary water to be used for higher-temperature circuits while still promoting condensation in the secondary heat exchanger.

You should familiarize yourself with these boilers because they provide a means of saving primary energy. Figure 6.8 shows a typical application. The condensing boiler is the lead boiler and will be operative throughout the heating season, making use of the latent heat from the flue gases in mild weather when the return temperature from the CVVT zones is low. The lag boiler would be a conventional boiler for use in severe weather with the standby facility of a further conventional boiler if required. Figure 6.8 shows CVVT control, which is achieved directly via the boiler rather than through a three-port control valve and actuator. It is therefore not possible to serve fan convectors or unit heaters from this system as they require constant temperature constant volume (CTCV) control.

Figure 6.9 shows multiple boilers: two conventional and one split condensing. There are two heating circuits, one with CVVT control via a three-way mixing valve and the other having CTCV control serving, say, fan convectors. The secondary heat exchanger in the split condensing boiler will increase the unit's thermal efficiency during a mild winter climate because of the effect of weather compensation at the mixing valve. The fan convectors operate at a constant temperature, and therefore return water is not low enough to be passed through the secondary heat exchanger. The high thermal efficiencies associated with condensing boilers are dependent upon the return water temperature being low (30–55 °C), and these units are therefore particularly appropriate for underfloor heating, swimming pool heating and systems employing low-temperature radiators.

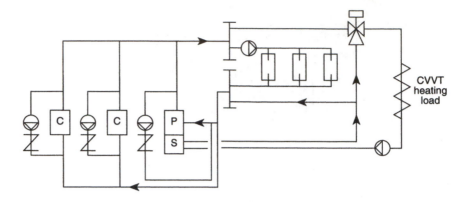

Figure 6.9 Two conventional and one split condensing boiler system: C, conventional boiler; P, primary heat exchanger; S, secondary heat exchanger of double condensing boiler. F&E tank, cold feed, open vent, isolating, regulating valves and control elements omitted.

TEMPERATURE CONTROLS FOR SPACE HEATING AND HOT WATER SUPPLY

Room temperature control

There are a number of space temperature control philosophies. For appliances such as radiators, the use of constant volume variable (circuit water) temperature control (CVVT) employing a three-port mixing valve is well known. In Figure 6.10a, the outdoor detector senses changes in climate, and the controller drives the actuator, thus repositioning the shoe in the valve. At design conditions there is no flow in the bypass, but as outdoor temperature rises the bypass port starts to open and the boiler port begins to close, thus allowing lower return temperature water to mix with the boiler water. The immersion thermostat senses the mixed water temperature and, if necessary, takes corrective action via the controller and valve actuator. A room thermostat or averaging thermostats may also be connected to the controls, although it is difficult to identify a suitable indoor location in practice (Figure 6.10a).

Single heat exchangers required to produce a constant temperature, such as air heater batteries and HWS calorifiers, are fitted with constant temperature variable volume (CTVV) control using a three-port mixing valve. Here the boiler water is diverted into the bypass as the air reaches the temperature setting of the duct thermostat located downstream in the supply air duct or in the return air duct (Figure 6.10b).

Note the location of the pump in each of Figure 6.10a and b. In Figure 6.10a the water flow through the boiler is interrupted when the control is on full bypass.

Constant temperature variable volume control is also achieved using two-port valves responding to a room thermostat and by using thermostatic radiator valves. In both cases, the water flow varies to maintain a constant room

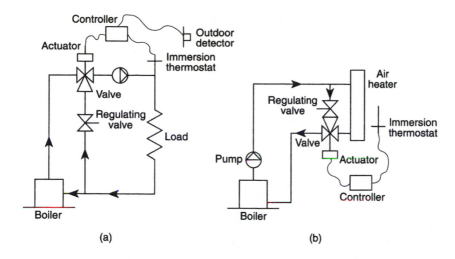

Figure 6.10   (a) CVVT control; (b) CTVV control.

temperature. It is important to consider the use of balanced pressure valves if the pump is dedicated to the circuit being controlled (see Chapter 3).

Space heating appliances such as unit heaters and fan convectors are controlled by a room thermostat wired in series with the fan motor. The heating circuit needs only the temperature control imposed by the boiler operating thermostat, thus providing constant temperature constant volume control (CTCV). When the fan is off, output is reduced by about 90%, and hot water is instantly available when the fan is reactivated by the room thermostat (Figure 6.11).

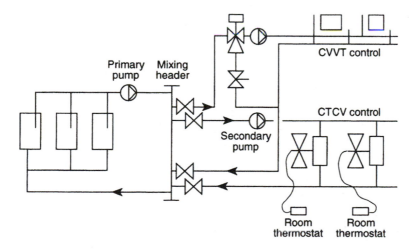

Figure 6.11   A simple schematic showing CVVT control to radiators and CTCV control to unit heaters.

Fan coil units on the other hand are controlled by three- or four-way control valves in conjunction with room sensors, the fans operating continuously during occupation of the room or building. The three-way mixing valve provides CTVV control.

Limitations on CVVT control

This type of control relies on compensating for changes in outdoor climate, and on its own does not give local control of heating appliances. It also suffers from the location of the outdoor detector, which, if located on the north wall of the building, will not compensate for the effects of solar heat gains or for that matter local indoor heat gains. However, CVVT control is to be recommended where appropriate as a means of reducing circuit flow temperatures in mild weather, thus offering energy savings. Local temperature control can be provided by employing, in addition, thermostatic radiator valves or two-port control valves, activated by a room thermostat to a group of radiators.

It may be appropriate to provide two or more zones of CVVT control when zoning the building (see Figures 6.2 and 6.12).

Temperature control for hot water supply

Figure 6.12 is a schematic showing a combined heating and hot water supply system. The CTVV primary circuit to the HWS calorifier would be controlled by an immersion thermostat located in the secondary water within the cylinder and set to 65 °C. The two CVVT heating zones may be required to offset the effects of building orientation and solar heat gain. If this provides adequate temperature control, it may not be necessary to fit TRVs to all the radiators. Remember, without them there is no *local* temperature control. Note that with the pump on the CTVV HWS primary circuit, there is constant water flow through the boiler plant and therefore no need for boiler return protection.

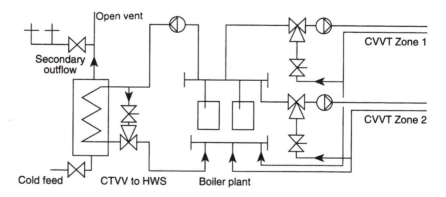

Figure 6.12 Illustrating 2 – CVVT and 1 – CTVV control circuit. Isolating valves, heating F&E pipe, open vent and F&E/CWS tanks omitted.

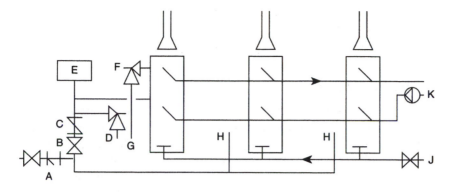

Figure 6.13   Three direct-fired HWS storage heaters showing direct MWS connections: A, MWS with isolating valve and strainer; B, pressure-limiting valve; C, check valve; D, expansion valve; E, expansion vessel; F, temperature/pressure relief valve; G, drains to tundish; H, repeat of items B–G to the two other heaters; J, gas supply; K, HWS flow and return.

There is a strong argument for separating the generation of HWS from the space heating plant on the grounds of efficiency. Direct-fired HWS generators connected directly to the rising main are widely used, and for larger rates of simultaneous flow can be connected in banks (see Figure 6.13).

SCHEMATIC AND LOGIC DIAGRAMS

You are now able to put together a schematic diagram showing space heating and hot water supply systems for a project (diagrams for indirect hot water supply are given in Chapters 8 and 9). It will include all the elements for each temperature control device. For example, CTVV control will include the valve body, valve actuator, immersion thermostat and controller. A logic diagram is a schematic, which has plant duties and main flow rates added: for example, boiler outputs, pump duties, HWS heater outputs and simultaneous flow rate, tank sizes, fan duties and heater battery duties.

BOILER TEMPERATURE CONTROL

Operating and limit thermostats

The operating thermostat is set to the design flow temperature, and the limit thermostat, which is wired in series, is set 5 K above the temperature of the operating stat. They both control the operation of the boiler fuel burner and are located in the boiler waterways. Control of boilers with modulating burners is more complex but the principle still applies. In the event of failure of both thermostats, the burner fires continuously and may cause the water to cavitate within the boiler or high-level pipes where supporting static pressure is at a minimum. This is the purpose of the open vent that is connected to the boiler as a safety device.

Variable-switching thermostat

A sensor takes the place of the operating thermostat and constantly monitors flow temperature, operating with a 5 K switching differential under heavy load. By measuring the angle of decay of the flow temperature the switching differential is widened automatically downwards with decreasing load. This reduces the frequency of boiler operation and brings down the system mean water temperature, thus reducing system output.

Early morning boost

If there is sufficient overload capacity (plant ratio F_3, see examples 1.8 and 1.9, Chapter 1) the boiler plant can be set to operate at an enhanced flow temperature on start-up, to heat the building more rapidly before occupation.

Frost protection

If the boiler plant has a shut-down period it may be necessary to ensure that the indoor temperature does not drop below about 10–12 °C and system water does not freeze. Frost protection overrides the time controls and can energize the pumps to circulate the system water as a first stage and subsequently operate the boiler plant.

Night setback

As an alternative to shutting the boiler plant down, room temperature controls can be set back from, say, 19–14 °C during part of the unoccupied period.

Fixed start/stop

Time controls having single or multiple channels with fixed start and stop periods for plant or pump operation are in common use. Plant would be timed to operate over the total heating period, and pumps or three-way valves would be timed to operate for different time-scheduled zones. Time controls are available for weekly, monthly or yearly time scheduling.

Optimum start/stop

For boiler plant requiring a regular weekly time schedule of, say, 5 days on and 2 days off, optimum start/stop control may be appropriate. The self-adaptive system stores and adjusts the start and stop times from the thermal response signature of the building to changes in outdoor climate in respect of indoor design temperature and prevailing indoor temperature. Preheat times are generally less than those for fixed start and are dependent upon the thermal response factor f_r. Some of the facilities offered for optimum start/stop are:

- optimum start/stop of boiler and pumps;
- weather compensation via boiler or three-port mixing valve;
- minimum boiler return temperature control;
- frost protection or night setback;
- day economization;
- multiple zone time scheduling via pumps or valves;
- optimum start boost;
- flue gas temperature monitor and alarm;
- oil level monitor and alarm;
- boiler sequencing;
- lead boiler changeover;
- boiler pump over-run;
- summer turnover for heating pumps.

Sequence control

Multiple boilers can be fired in a sequence dependent upon the prevailing heat load such that on light loads only one boiler is operational, with the other boilers becoming operational sequentially as the load increases.

If the boilers each have a dedicated pump they must operate for a short period after the burner has shut down in response to a fall-off in system load to dissipate the heat in the combustion chamber; otherwise cavitation can occur within the boiler waterways.

Boiler return control

This matter is discussed earlier. Refer to Figures 6.6 and 6.7a and b.

Zoning: time scheduling

This subject is considered in Figure 6.1 and under 'Fixed start/stop'.

CONTROL SYSTEMS

Traditionally many building services systems are controlled using either pneumatics or electric/electronic and mechanical devices such as the five elements in CVVT control: valve body, valve actuator, immersion thermostat,

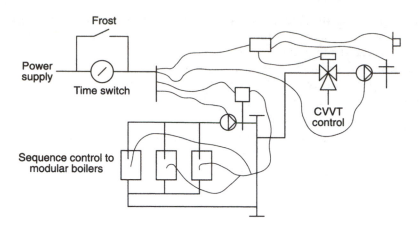

Figure 6.14 Electromechanical system controls.

outdoor detector and controller. These may be wired back to a central control panel in which is located the time-scheduling device and frost protection (see Figure 6.14). This type of control still has its place, albeit in an increasingly limited way in commercial and industrial installations.

6.3   Building energy management systems

Direct digital control and supervisory control can be more user-friendly and can give the user more control over the building services systems either locally or remotely via a modem to a *building energy management system* (BEMS). The capital costs and advantages of a BEMS depend upon whether the user has the time and commitment to use this facility and take full advantage of the technology.

BEMS is the subject of another book by the publishers. Here it is therefore only necessary to introduce the concept. The *local area network* (LAN) might include BEMS outstations or original equipment manufacturers' (OEM) outstations, central station and printer. This would be linked to a modem if the final control and monitoring location is remote, say, in another building some miles away. Software is generated and dedicated to operate the controls and relay system conditions such as temperature, relative humidity, pressure, pressure drop, and status such as duty plant operation, standby plant operation. These conditions can be called up on a visual display unit (VDU) or monitor, and will include system logic diagrams. The way a BEMS is connected to a LAN is called the *topology*, of which there are basically three, bus, star and ring. Figure 6.15 illustrates the principles of a BEMS using a *bus topology*.

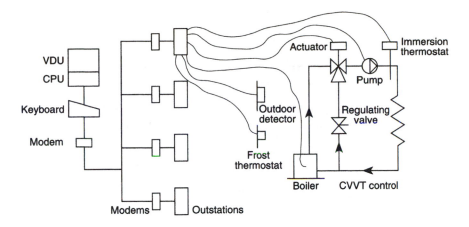

Figure 6.15 Principal features of a BEMS with a bus topology and remote central station.

The keyboard and central processing unit (CPU) complete with visual display unit (VDU), collectively called a *central station*, are connected to the LAN via a modem. The example in Figure 6.15 shows a LAN consisting of four outstations, each with a modem connected on a bus topology via a modem to a central station. One outstation is shown providing CVVT control, boiler and pump control and frost protection to a circuit of radiators, for illustration purposes.

There are two other topologies in use with BEMS. They are *star* and *ring*, and are illustrated in Figure 6.16a and b. With the star network, outstations can communicate to each other only through the central station, whereas

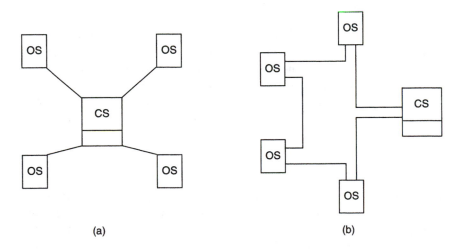

Figure 6.16 (a) Star network; (b) ring network. CS, central station; OS, outstation.

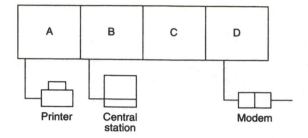

Figure 6.17 Single packaged local station: A, plant and sensor in/out; B, control and monitoring processing; C, user interface and programming; D, data archival remote communication.

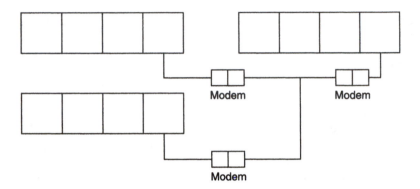

Figure 6.18 Networked local stations.

with the ring network, outstations can communicate independently of the central station.

There are two other protocols that are worth mentioning here for the smaller building services systems. The *single packaged local station* shown in Figure 6.17 is suitable for the small commercial/industrial installation. There is no reason, however, why the single station cannot be linked following expansion and building extension. With the arrangement in Figure 6.18, each local station can communicate following expansion of the enterprise.

OUTSTATIONS

There are two types: the dumb outstation and the intelligent outstation. The *dumb outstation* has the executive control at the central station. This is called *centralized intelligence*. The *intelligent outstation* takes control without input from the supervisor station or central station. This is termed *distributed intelligence* (see Figure 6.19).

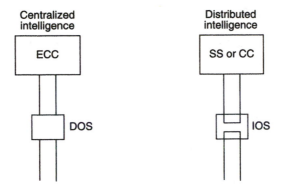

Centralized intelligence	Distributed intelligence
ECC	SS or CC
DOS	IOS

Figure 6.19   The dumb and the intelligent outstation: DOS, dumb outstation; ECC, executive central control; IOS, intelligent outstation; SS or CC, supervisor station or central control.

OUTSTATION FUNCTIONS

These are split into three levels:

1. *High level*: remote communication, user interface, optimizer control, cascade control, maintain trend logs, maintain event logs.
2. *Mid level*: proportional plus integral plus differential (PID) control, main data (MD) control, alarm check, alarm communication, program-defined interlocks, calendar/clock control, linearize measurement, convert to engineering units.
3. *Low level*: hard-wired interlocks, elapse timer control, scale measurements, check input limits, debounce inputs, count pulses, plant interface, drive outputs, sensor interface, scan inputs.

INTERLOCK SYSTEMS

Interlocks can be considered as 'don't till' statements, and are of importance in defining the control strategy and in detailing the schematic.

An example of a system of interlocks at the commencement of plant operation might be

- *Don't* start primary heating pump *till* the time is right.
- *Don't* start the lead boiler *unless* primary pump is energized.
- *Don't* start secondary pump on zone 1 *till* boiler primary circuit is at 80 °C.
- *Don't* start secondary pump on zone 2 *till* zone 1 is at 80 °C.
- *Don't* start unit heater fans on zone 3 *till* the circuit water is at 80 °C.

SUPERVISOR STATION AND CENTRAL STATION FUNCTIONS

These can include the following functions in addition to those listed for the outstation:

- plant supervision;
- maintenance supervision;
- security supervision;
- energy monitoring;
- environmental monitoring;
- system development;
- plant executive control;
- reporting;
- data archival;
- design evaluation.

SUPERVISORY AND DIRECT DIGITAL CONTROL

There is a difference between supervisory control and direct digital control, although the former is often called by the latter name. The supervisory system uses the local controller, whereas direct digital control (DDC) dispenses with it (see Figures 6.20 and 6.21).

6.4   Control strategies for heating systems

Recommendations have been made relating to controls for heating systems. These include the following:

1. The boiler primary circuit should be pumped at constant volume and be hydraulically independent of the secondary circuits.
2. Domestic hot water should be provided by a separate system.

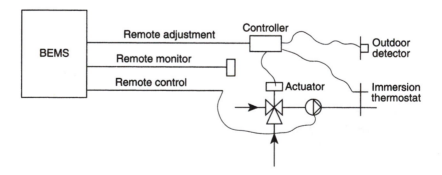

Figure 6.20   A supervisory system providing CVVT control.

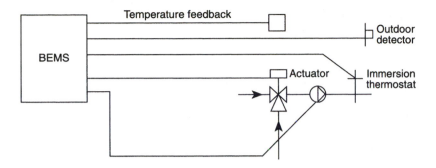

Figure 6.21   Direct digital control system providing CVVT control.

3. Systems over 30 kW should be controlled by an optimizer.
4. The temperature to each zone should be compensated according to outdoor temperature (clearly this cannot apply to fan convectors or unit heaters).
5. The zones themselves should be selected on the basis of:

    (a) solar heat gain (building orientation);
    (b) building exposure (multi-storey buildings – horizontal zoning);
    (c) occupancy times;
    (d) building thermal response;
    (e) types of heating appliance.

6. Space temperature reset of the space heating controls.
7. Frost protection during off periods.

Recommendations have been made for energy-saving features on heating system controllers, and they include:

1. optimizers for loads over 30 kW;
2. time clocks for loads below 30 kW;
3. compensated flow control of the boilers (clearly not possible for fan convectors or unit heaters);
4. protection against frost;
5. sequence control of multiple boilers;
6. control of demand;
7. heating pump run-on facility;
8. summer exercising of pumps;
9. time delays to prevent boilers starting with sudden temporary demand;
10. flue gas monitoring and alarm;
11. 7-day programming;
12. holiday programming;
13. auto summer/winter time change;
14. graphic display of operating times and off times;

15.   minimum run times;
16.   maximum heating flow temperature.

SYSTEM OPERATION METHOD STATEMENT

An important part of the schematic or logic diagram is a written statement that explains how you intend the systems to operate. The two go hand in hand, and the method statement may indeed help to refine the logic diagram. It will also clarify how, for example, you want the boilers to operate, and may lead to adjustments to the schematic. It is important therefore that both are given time for development in the early stages of project design.

The *Method Statement* will include:

1.   details of zoning by time scheduling and control of temperature;
2.   description of plant and circuits, which would include space heating, ventilation, hot water supply and cold water supply;
3.   description of interlocks;
4.   client interface (how the client can use the systems);
5.   specialist interface (where the specialist must be called in to adjust/ monitor/refine and maintain).

Case studies

1.   Figure 6.22 shows a schematic of a space heating system. It includes a zone of radiators providing CVVT control to offices and background heating to a conference room, a CTCV circuit serving unit heaters to a workshop, and CTVV circuit to an air handling unit that provides tempered air to the conference room when it is in use. Write an operating method statement for the schematic and suggest what, if any, additions you would make to the control protocol.
2.   A multi-storey office block on a north–south axis is for a client who will want it partitioned into offices. It has a central corridor with a staircase at each end. Each floor will be let independently. Write a system operation method statement for space heating and hot water services in this building.
3.   A college campus consists of five buildings, offering the following facilities: building crafts, engineering workshops, catering and restaurant, business studies and college administration, and sports hall.

There are four plant rooms, with the building craft and engineering workshop buildings having a common plant room and the

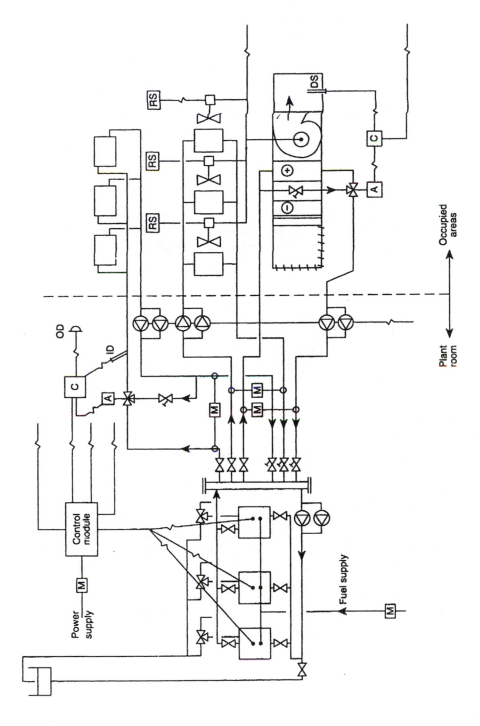

Figure 6.22 Case Study 6.1: schematic diagram for the energy survey.

remaining buildings having independent plant. A BEMS system is specified for the campus. Draw a schematic of the plant rooms and example heating, HWS and mechanical ventilation systems in each, connected via LANs to a central control station.

4. Describe the protocols for linking local area networks in a building energy management system and identify their strengths and weaknesses.

5. A client wants to update a space heating plant, which is connected to four radiator zones, each controlled by a three-port mixing valve providing CVVT. You decide to use two modular boilers and one condensing boiler, which you hope will give an energy benefit, particularly at the beginning and at the end of the heating season. Draw a schematic for the boiler plant and explain how the condensing boiler can take advantage of zone return temperatures below 60 °C. Now write a plant operation method statement and check that what you intended in the schematic will actually occur.

6.5   Further reading

1. CIBSE Guide H: Building control systems
2. CIBSE Guide B1 Heating
3. Building Services OPUS Design File for manufacturers' products
4. Building energy management systems Spon Press

6.6   Chapter closure

Following the topics introduced in this chapter you are now able to recommend temperature controls for a variety of space heaters. You can prepare schematic diagrams for multi-circuit systems, showing appropriate temperature, boiler and bypass controls and connections to plant. You are in a position to select and propose boiler temperature control options. You are cognizant of the recommended controls on heating systems and the recommended energy-saving features and able to select those that are appropriate. You can write a system operation method statement. Finally, you have a working knowledge of the principles of supervisory and direct digital control offered in building energy management systems.

It is possible to study this chapter without recourse to manufacturers' literature. However, it is strongly recommended that you investigate the market for boilers and controls.

The application of probability and demand units in design 7

Nomenclature

Cumulative P_m	probability of satisfying the demand
CWS	cold water service
DU	demand unit
HWS	hot water service
K	constant
m	number of fittings simultaneously discharging
MWS	mains water service
n	total number of fittings in a system
P	probability or usage ratio
P_m	probability of occurrence
SD	standard deviation
T	average time between occasions of use
t	average time the draw-off discharges for each occasion of use
X	the mean

7.1   Introduction

Sizing hot water and cold water supply systems, like space heating systems, is done from a knowledge of flow rates in each pipe section. However, determination of the flow rates for hot and cold water supply is a function of the simultaneous consumption of water. This will vary from one system to the next, depending upon the usage of the draw-off points or fittings on the system, for it would not normally be appropriate to assume that *all* draw-off points in that system will be in use simultaneously.

Simultaneous flow may last only for a matter of a few seconds. It will not necessarily occur from the *same* draw-off points in a system on each occasion of use. There is one application, however, when it is essential to assume that all the draw-off points will be in use simultaneously. This will be mentioned later.

Apart from this and one or two other exceptions, therefore, the concept of *probability* or *usage ratio P* must be introduced into the determination of simultaneous flow before pipe sizing can be considered, otherwise the pipework will be oversized and the system would therefore be unnecessarily costly to install.

7.2 Probability or usage ratio *P*

P is defined as t/T, where t is the average time for which the draw-off point is discharging for each occasion of use, and T is the average time between occasions of use. The *CIBSE Guide* lists values for usage ratio P for different sanitary appliances: they range from 0.014 to 0.448.

If for a fitting $t = 60\,\text{s}$ and $T = 300\,\text{s}$ then for that fitting the usage ratio $P = 60/300 = 0.2$. If, exceptionally, the fitting is used continuously for five occasions, $P = (60 \times 5)/300 = 1.0$. That is to say, the fitting is in continuous use. If, exceptionally, a *group* of fittings is being used simultaneously for, say, 60 s, then clearly the total flow when all the fittings are discharging must be used to size the associated pipes even if these fittings are not in use for some time after the event. Clearly the usage ratio is applied only to fittings not in use continuously when $P < 1$.

THE BINOMIAL DISTRIBUTION

Probability P follows the binomial distribution, and for hot and cold water supply:

$$P_m = \frac{n!}{m!(n-m)!} \times P^m (1 - P)^{n-m}$$

where P_m is the probability of occurrence; cumulative P_m is the probability of satisfying the demand; n is the total number of fittings having the same probability; and m is the number of fittings in use at any one time. $n!$ is called factorial n and stands for the product of whole numbers from n down to 1. For example, using this notation, factorial four $= 4! = 4 \times 3 \times 2 \times 1 = 24$.

Note: When $m = 0$, $P_m = (1 - P)^n$; when $m = n$, $P_m = P^n$. Furthermore, $0! = 1$ and $1! = 1$. You should now confirm these notes relating to the binomial distribution.

Case Study 7.1

Ten HWS draw-off points are installed in a building and connected to a common secondary outflow. Each point has a usage ratio P of 0.3. Adopt the binomial distribution and tabulate the probability of occurrence and the probability of satisfying the demand. Draw a probability distribution histogram and calculate the mean and standard deviation. Show that the binomial distribution can be simplified to the formula below for the simultaneous number of points (m) discharging for a given common probability when the probability of satisfying the demand is 99.8%:

$$m \approx nP + 1.8(2nP(1 - P))^{0.5}$$

SOLUTION

P_m and cumulative P_m are tabulated in Table 7.1. The number of draw-off points in use simultaneously, assuming for example a 98% probability of satisfying the demand, is, from the tabulated solution, six, at 98.9%. The probability of the event occurring is 3.7%. You will notice that the highest probability that an event will occur is 26.7% for three draw-off points and not for less or more than three. The probability distribution histogram is drawn from the values of P_m and the number of draw-off points in use, m (Figure 7.1).

The mean and standard deviation are two measures that describe the frequency distribution of demand. The *mean*, X, is a measure of the

Table 7.1 Case Study 7.1: P_m and cumulative P_m

Number in use, m	Binomial distribution	P_m	Cumulative P_m
0	$(1 - 0.3)^{10}$	0.028	0.028
1	$(10!/1!\,9!)(0.3)^{1}(1 - 0.3)^{9}$	0.121	0.149
2	$(10!/2!\,8!)(0.3)^{2}(1 - 0.3)^{8}$	0.233	0.382
3	$(10!/3!\,7!)(0.3)^{3}(1 - 0.3)^{7}$	**0.267**	0.649
4	$(10!/4!\,6!)(0.3)^{4}(1 - 0.3)^{6}$	0.2	0.849
5	$(10!/5!\,5!)(0.3)^{5}(1 - 0.3)^{5}$	0.103	0.952
6	$(10!/6!\,4!)(0.3)^{6}(1 - 0.3)^{4}$	**0.037**	**0.989**
7	$(10!/7!\,3!)(0.3)^{7}(1 - 0.3)^{3}$	0.009	**0.998**
8	$(10!/8!\,2!)(0.3)^{8}(1 - 0.3)^{2}$	0.001	0.999
9	$(10!/9!\,1!)(0.3)^{9}(1 - 0.3)^{1}$	0.000138	0.999
10	$(0.3)^{10}$	0.0000059	0.999

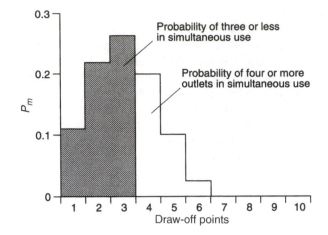

Figure 7.1 Case Study 7.1: the histogram of probability distribution of demand.

central tendency of the distribution, and the *standard deviation*, SD, is a measure of the spread of the distribution:

$$X = \frac{\sum fm}{\sum f}$$

and

$$SD = \left(\frac{\sum fm^2}{\sum f} - X^2\right)^{0.5}$$

where for convenience $P_m = f$.

The values in these formulae, obtained after multiplying $f(P_m)$ by 1000 to remove the decimal place, are listed in Table 7.2. Applying the formulae for X and SD:

$$X = \frac{2996}{999} = 2.99$$

and

$$SD = \left(\frac{11\,068}{999} - 8.994\right)^{0.5} = 1.44$$

If the number of draw-off points, m, varies discretely, i.e. increases by one whole number at a time, and the minimum number is zero and the maximum number is finite, then the distribution is binomial and allows a simplification of the formula such that:

$$X = nP = 10 \times 0.3 = 3.0$$

Table 7.2 Case Study 7.1

f	m	fm	fm²
28	0	0	0
121	1	121	121
233	2	466	932
267	3	801	2403
200	4	800	3200
103	5	515	2575
37	6	222	1332
9	7	63	441
1	8	8	64
0	9	0	0
0	10	0	0
Total: 999		2996	11 068

and

$$\text{SD} = (nP(1 - P))^{0.5} = (10 \times 0.3 \times 0.7)^{0.5} = 1.45$$

Clearly the error in evaluating X and SD is insignificant, and therefore the binomial formula may be written in its simplified form as: the number of draw-off points simultaneously discharging, m, is equal to the mean plus a constant K times the standard deviation. This is a mathematical statement that you may know. Substituting for X and SD,

$$m = nP + K(nP(1 - P))^{0.5}$$

The constant K is dependent upon the probability of satisfying the demand, P_m, and for a 99.8% probability, $K = 2.54$ taken from Table 7.3. Thus

$$m = nP + 2.54(nP(1 - P))^{0.5}$$

The simplified formula is more commonly written as

$$m = nP + 1.8(2nP(1 - P))^{0.5} \tag{7.1}$$

Table 7.3

Values of P_m	Constant K
0.8	0.842
0.9	1.281
0.95	1.645
0.99	2.326
0.998	2.54
0.999	3.09

Do the two formulae for m (the number of fittings simultaneously discharging) agree?

Applying this formula to the solution to the number of points simultaneously discharging,

$$m = 10 \times 0.3 + 1.8(2 \times 10 \times 0.3 \times 0.7)^{0.5} = 7$$

This solution is appropriate for a 99.8% probability of satisfying the demand. Check with the first set of data in Table 7.1 with the solution to this case study. Does it agree? This demonstrates that the simplified formula is valid for use in the determination of the number of draw-off points simultaneously discharging for a common value of probability.

Case Study 7.2

Determine the usage ratio P for bathrooms in which 5 min may be taken as the time to fill a bath, after which there is a period of 20 min before the bath might again require filling. Substitute into equation (7.1) and determine the simultaneous demand for 20 baths each having a flow rate of 0.5 litres/s. Determine the simultaneous demand when the total number of baths connected to a common outflow is 20 and the discharge from each is 0.5 litres/s for a common usage ratio of $P = 0.1$.

SOLUTION

The first usage ratio $P = 5/(20 + 5) = 0.2$.
Substituting into equation (7.1),

$$m = 0.2n + 1.8(2n \times 0.2 \times 0.8)^{0.5}$$

This reduces the equation to

$$m = 0.2n + (n)^{0.5}$$

Thus for a usage ratio of $P = 0.2$,

$$m = 0.2n + (n)^{0.5} \tag{7.2}$$

Substituting $n = 20$,

$$m = 0.2 \times 20 + (20)^{0.5}$$

from which

$$m = 8$$

and the simultaneous demand is equal to

$$\frac{m}{n} \times \text{total flow} = \frac{8}{20} \times 20 \times 0.5 = 4\,\text{litres/s}$$

If all the baths were in use together,

$$\text{flow} = 20 \times 0.5 = 10\,\text{litres/s}$$

This clearly demonstrates the effect of the usage ratio P and its effect on pipe size.

For the second part of the case study the usage ratio $P = 0.1$. Substituting this value for P into equation (7.1),

$$m = 20 \times 0.1 + 1.8(2 \times 20 \times 0.1 \times 0.9)^{0.5}$$

from which

$$m = 5$$

and the simultaneous demand is equal to

$$5 \times 0.5 = 2.5\,\text{litres/s}$$

or using

$$\frac{m}{n} \times \text{total flow} = \frac{5}{20} \times 20 \times 0.5 = 2.5\,\text{litres/s}$$

These results are summarized in Table 7.4.

CONCLUSION

The effect of the usage ratio P on simultaneous flow compared with total flow is clearly apparent. If the number of baths is doubled the simultaneous flow is not, and for $P = 0.2$ the simultaneous flow for 40 baths is calculated as 7 litres/s. For $P = 0.1$ the simultaneous flow is 4.5 litres/s. Do you agree?

As the total number of similar fittings (with a common value of P) in a system increases, therefore, the simultaneous flow does not increase

Table 7.4 Case Study 7.2: summary of results

P	n	m	Simultaneous flow	Total flow
0.2	20	8	4	10
0.1	20	5	2.5	10

at the same rate. Note also that in the solutions for m the numerical value is taken to its nearest whole number.

Case Study 7.3

Figure 7.2 shows a centralized HWS system diagrammatically in elevation. Determine the simultaneous flow in each pipe section if the draw-off points have the following common usage ratios: (a) when $P = 0.2$ and (b) when $P = 0.4$.

Each floor has the following fittings: two baths, six basins, one sink and two showers. The recommended discharge for each fitting is: bath = 0.4 litres/s, basin = 0.15 litres/s, sink = 0.3 litres/s and shower = 0.15 litres/s.

SOLUTION

The total flow to each floor is calculated to be 2.3 litres/s, and for seven floors the total flow will be 16.1 litres/s.

(a) Table 7.5a lists the results when $P = 0.2$ and from equation (7.2), $m = 0.2n + (n)^{0.5}$.

(b) Table 7.5b lists the results when $P = 0.4$ and from equation (7.1), $m = nP + 1.8(2nP(1 - P))^{0.5}$.

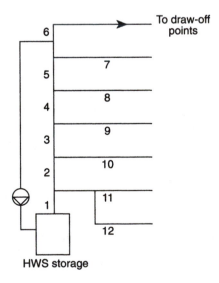

Figure 7.2 Centralized hot water service to seven floors (open vent, cold feed and CWS tank omitted).

Table 7.5 Case Study 7.3: results

Section	1	2	3	4	5	6	7	8	9	10	11	12
(a)												
Total flow	16.1	11.5	9.2	6.9	4.6	2.3	2.3	2.3	2.3	2.3	2.3	2.3
n	77	55	44	33	22	11	11	11	11	11	11	11
m	24	18	15	12	9	6	6	6	6	6	6	6
m/n	0.31	0.33	0.34	0.36	0.41	0.55	0.55	0.55	0.55	0.55	0.55	0.55
Simultaneous flow (litres/s)	5.0	3.8	3.13	2.5	1.9	1.27	1.27	1.27	1.27	1.27	1.27	1.27
(b)												
m	42	31	26	20	15	9	9	9	9	9	9	9
m/n	0.55	0.56	0.59	0.61	0.68	0.82	0.82	0.82	0.82	0.82	0.82	0.82
Simultaneous flow (litres/s)	8.86	6.44	5.43	4.21	3.13	1.89	1.89	1.89	1.89	1.89	1.89	1.89

CONCLUSION

The simultaneous flows in each pipe section should first of all be compared with the total flow. The simultaneous flows for (a) and (b) should then be compared. These comparisons will give you a feel for the effects of the usage ratio on total flow rates and when the ratio itself is varied.

MULTIPLE PROBABILITIES

The application of the usage ratio *P* is straightforward enough when the same value is adopted for all the draw-off points in a system. However, as indicated at the beginning of the chapter, the usage ratio varies with the type of fitting in a system of hot or cold water supply, and in such circumstances it is not possible to determine simultaneous flow without initially reducing the fittings on a system to a common value for *P*.

Consider the next case study.

Case Study 7.4

Determine the simultaneous outflow from a hot water service storage vessel serving the following system:

Type of draw-off point	A	B
Number of draw-off points	40	20
Discharge (litres/s)	0.2	1
Usage ratio, *P*	0.2	0.3

SOLUTION

As equation (7.2) is easier to manipulate, it is convenient to convert all the draw-off points in the system to an equivalent usage ratio of $P = 0.2$.
    Consider B-type fitting. Adopting equation (7.1),

$$m = 20 \times 0.3 + 1.8(2 \times 20 \times 0.3 \times 0.7)^{0.5} = 11$$

This is equivalent to n points at $P = 0.2$, and from equation (7.2),

$$m = 0.2n + (n)^{0.5}$$

Substituting,

$$11 = 0.2n + (n)^{0.5}$$

If $y^2 = n$, then

$$11 = 0.2y^2 + y$$

and

$$0 = 0.2y^2 + y - 11$$

Adopting the quadratic formula:

$$y = \frac{-1 \pm (b^2 - 4ac)^{0.5}}{2a}$$

and substituting,

$$y = \frac{-1 \pm (1 + (4 \times 0.2 \times 11))^{0.5}}{2 \times 0.2}$$

from which the positive solution is $y = 5.3$, and therefore as $n = y^2$, $n = 28$. Thus 28 type B fittings at $P = 0.2$ are equivalent to 20 type B draw-off points at $P = 0.3$.
    The *equivalent* system having $P = 0.2$ is therefore

A	40 points at 0.2 litres/s =	8 litres/s
B	28 points at 1.0 litres/s =	28 litres/s
Total	68	36 litres/s

The number of points simultaneously discharging will be, adopting equation (7.2),

$$m = 0.2n + (n)^{0.5}$$

Substituting

$$m = 0.2 \times 68 + (68)^{0.5} = 22$$

Simultaneous demand from the vessel will be

$$\frac{m}{n} \times \text{total flow} = \frac{22}{68} \times 36 = 11.6 \, \text{litres/s}$$

Case Study 7.5

Figure 7.3 shows a cold water down service to a number of fittings from a high-level storage tank. Determine from the data the simultaneous flows in each pipe section.

DATA

Branch	A	B	C	D
N	50	20	30	80
Discharge	0.5	0.15	0.3	0.4 litres/s each
P	0.4	0.3	0.2	0.5

SOLUTION

There is again an opportunity to adopt the simplified equation (7.2) if branches A, B and D are reduced to a common probability P of 0.2, which is the usage ratio for branch C.

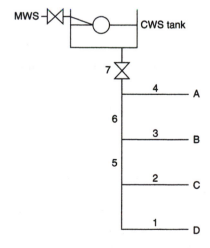

Figure 7.3   Cold water down service to fittings having more than one value of P.

Converting type A fittings to an equivalent number of fittings having a usage ratio of $P = 0.2$, from equation (7.1):

$$m = (50 \times 0.4) + 1.8(2 \times 50 \times 0.4 \times 0.6)^{0.5}$$

from which

$$m = 29$$

This is equivalent to n points having a probability of $P = 0.2$. From equation (7.2),

$$29 = 0.2n + (n)^{0.5}$$

If $y^2 = n$ then

$$0 = 0.2y^2 + y - 29$$

Adopting the quadratic formula,

$$y = \frac{-1 \pm (1 + (4 \times 0.2 \times 29))^{0.5}}{0.4}$$

from which

$$y = 9.8 \quad \text{and, therefore} \quad n = 96$$

Thus 50 type A fittings having a probability of 0.4 are equivalent to 96 fittings of the same type having a probability of 0.2.

Converting type B fittings to an equivalent number of fittings having a probability $P = 0.2$, from equation (7.1),

$$m = (20 \times 0.3) + 1.8(2 \times 20 \times 0.3 \times 0.7)^{0.5}$$

from which

$$m = 11$$

This is equivalent to n points having a probability of $P = 0.2$. From equation (7.2),

$$11 = 0.2n + (n)^{0.5}$$

If $y^2 = n$, then

$$0 = 0.2y^2 + y - 11$$

and using the quadratic formula

$$y = 5.32 \quad \text{and, therefore} \quad n = (5.32)^2 = 28$$

Thus 20 type B fittings having a probability of 0.3 are equivalent to 28 similar fittings having a probability of 0.2.

Converting type D fittings to an equivalent number having a probability $P = 0.2$, from equation (7.1),

$$m = (80 \times 0.5) + 1.8(2 \times 80 \times 0.5 \times 0.5)^{0.5}$$

from which

$$m = 51$$

This is equivalent to n points having a probability of $P = 0.2$. From equation (7.2),

$$51 = 0.2n + (n)^{0.5}$$

If $y^2 = n$ then

$$0 = 0.2y^2 + y - 51$$

and from the quadratic formula

$$y = 13.66 \quad \text{and, therefore} \quad n = (13.66)^2 = 187$$

Thus 80 type D fittings are equivalent to 187 similar fittings having a probability of 0.2.

All the fittings in the cold water down service system can now be expressed in terms of an equivalent system of fittings having a common probability. This allows the simultaneous flows in each pipe section to be added back to the common outflow at the tank. It is not possible to add simultaneous flows resulting from different usage ratios because distribution pipes (sections 5, 6 and 7 in this case study) would have to handle simultaneous flows of differing probabilities.

The solutions based upon the common probability for the system of $P = 0.2$ are listed in Table 7.6. The determination of m for distribution pipes 5, 6 and 7 highlighted in the tabulation must be done, and a sample calculation is given below for pipe section 5:

187 equivalent fittings at 0.4 litres/s = 74.8 litres/s
 30 equivalent fittings at 0.3 litres/s = 9.0 litres/s
217 total   83.8 litres/s total

Table 7.6 Case Study 7.5

Section	1	2	3	4	5	6	7	
n	187	30	28	96	217	245	341	
m	51	11	11	29	**58**	**65**	**87**	
Total flow (litres/s)	74.8	9	4.2	48	83.8	88	136	
m/n	0.27	0.37	0.39	0.3	0.27	0.26	0.25	
$m/n \times$ total flow	20.4	3.3	1.65	14.5	22.4	23.3	34.7	simultaneous flows (litres/s)

Thus for a common probability of 0.2, from equation (7.2),

$$m = (0.2 \times 217) + (217)^{0.5}$$

from which

$$m = 58$$

Do you agree with the determination of m for distribution pipes 6 and 7?

Case Study 7.6

Figure 7.4 shows a system of boosted cold water supply serving four branches, each having a discrete usage ratio. Determine the simultaneous flow of water in each pipe section for the purposes of pipe sizing.

DATA

Pipe branch	A	B	C	D
n	25	30	20	20
Discharge	0.15	0.3	0.2	0.15 litres/s each
P	0.01	0.1	0.05	0.01

SOLUTION

It is possible to adopt the simplified equation (7.2) if the system is reduced to the common probability of 0.2. Alternatively, as a usage ratio of 0.01 appears twice in the system, it is equally possible and possibly quicker to reduce the system to $P = 0.01$, in which case equation (7.1) must be exclusively adopted. Thus

$$m = 0.01n + 1.8(2 \times n \times 0.01 \times 0.99)^{0.5}$$

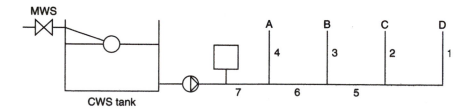

Figure 7.4   Boosted water system with multiple usage ratios.

Putting this in quadratic form:

$$m = 0.01n + 1.8(n)^{0.5} \times (0.0198)^{0.5}$$

from which

$$m = 0.01n + 0.253(n)^{0.5}$$

and if $y^2 = n$ then

$$0 = 0.01y^2 + 0.253y - m$$

Considering type A fittings:

$$m = (0.01 \times 25) + 0.253(25)^{0.5}$$

from which

$$m = 2$$

Considering type D fittings:

$$m = (0.01 \times 20) + 0.253(20)^{0.5}$$

from which

$$m = 1$$

Considering type B fittings, equation (7.1) in its original form applies:

$$m = (30 \times 0.1) + 1.8(2 \times 30 \times 0.1 \times 0.9)^{0.5}$$

from which

$$m = 7$$

This is equivalent to *n* fittings having a $P = 0.01$. Thus by substitution,

$$0 = 0.01y^2 + 0.253y - 7$$

Adopting the quadratic formula, $y = 16.67$ and therefore $n = 278$. Thus 30 B type fittings are equivalent to 278 fittings having a $P = 0.01$.
   Considering type C fittings, equation (7.1) in its original form applies:

$$m = (20 \times 0.05) + 1.8(2 \times 20 \times 0.05 \times 0.95)^{0.5}$$

from which

$$m = 3$$

Table 7.7   Case Study 7.6

Section	1	2	3	4	5	6	7	
n	20	77	278	25	97	375	400	
m	1	3	7	2	3	9	9	
Total flow (litres/s)	3	15.4	83.4	3.75	18.4	101.8	105.6	
m/n	0.05	0.04	0.025	0.08	0.031	0.024	0.0225	
m/n × total flow	0.15	0.6	2.1	0.3	0.57	2.44	2.38	simultaneous flows (litres/s)

This is equivalent to n fittings having a $P = 0.01$. Thus by substitution,

$$0 = 0.01y^2 + 0.253y - 3$$

Adopting the quadratic formula, $y = 8.79$ and therefore $n = 77$. Thus 20 type C fittings are equivalent to 77 fittings having a $P = 0.01$. These solutions are listed in Table 7.7.

As with Case Study 7.5, values of m must be determined for distribution pipes 5, 6 and 7. Do you agree with the tabulated solutions?

You are now invited to undertake the solution to this case study by reducing the system to the common probability of 0.2. You can then see whether the simultaneous flow rates you have calculated for each pipe section agree with the solution above. They may not agree exactly; apart from any errors, can you explain why?

7.3   The system of demand units (DU)

This is not to be confused with discharge units, which are used in the calculation of simultaneous flow for drainage systems. Clearly, from the work in Case Studies 7.4, 7.5 and 7.6, systems having fittings with more than one usage ratio involve tedious calculations, which at best could be transposed to a spreadsheet or database.

The system of demand units simplifies the whole process of determining the simultaneous flow in each pipe section for hot or cold water supply system having more than one usage ratio P. However, it is based upon a maximum probability of $P = 0.1$, and this imposes limits upon the wider application of demand units. It is important therefore to satisfy oneself that a proposed design for hot or cold water supply can be reconciled with the application of demand units where P is not greater than 0.1, otherwise the approach detailed in Case Studies 7.1–7.6 should be adopted.

Table 7.8 lists scales of practical demand units for some typical fittings. The types of application in the table can be misleading, as they refer to the value of the usage ratio, which for *congested* is equivalent to $P = 0.1$. The

Table 7.8 Scales of demand units

Fitting	Type of application		
	Congested T = 5 min	Public T = 10 min	Private T = 20 min
Basin	10	5	3
Bath	47	25	12
Sink	43	22	11
WC	22	10	5
Dishwasher	60	30	20
Clothes washer	40	20	10

public and private applications give rise to usage ratios progressively lower than $P = 0.1$.

Having identified the level of use for the various fittings in the system, each can be assigned its practical demand units, and these can be totalled back, section by pipe section, to the common outflow pipe. Table 7.9 is then used to translate the demand units into flow rates.

You will notice that, for example, 200 DU gives a flow of 0.8 litres/s; 400 DU gives a flow of 1.3 litres/s. In other words there is not a doubling of flow, and as the number of fittings increases, the simultaneous flow rises by smaller increments. This is also apparent if a system is designed by adopting usage ratios.

Case Study 7.7

Figure 7.5 shows a centralized HWS installation in elevation. Each branch has the following fittings and demand unit application: one sink (congested), two showers (congested), five basins (public), five basins (private), three baths (congested). Determine the simultaneous flow in each pipe section.

SOLUTION

From Table 7.8, it is relatively easy to determine the total number of DUs at each branch: one sink at 43, two showers at 10, five basins at 5, five basins at 3 and three baths at 47, giving a total of 244 DU. It is then appropriate to tabulate the solutions (Table 7.10). The total flow rates are given to show the effect of the application of the scale of demand units, and are obtained from the recommended discharge rates for fittings given in Case Study 7.3.

Table 7.9 Conversion of demand units to flow rates

Demand units									*Simultaneous flow* (litres/s)											
	0	50	100	150	200	250	300	350	400	450	500	550	600	650	700	750	800	850	900	950
0	0	0.3	0.5	0.6	0.8	0.9	1.0	1.2	1.3	1.4	1.5	1.6	1.7	1.9	2.0	2.1	2.2	2.3	2.4	2.5
1000	2.6	2.7	2.8	2.9	3.0	3.1	3.2	3.3	3.4	3.5	3.6	3.7	3.8	3.9	4.0	4.1	4.2	4.3	4.4	4.5
2000	4.6	4.7	4.8	4.9	5.0	5.1	5.1	5.2	5.3	5.4	5.5	5.6	5.7	5.8	5.9	6.0	6.1	6.2	6.3	6.4

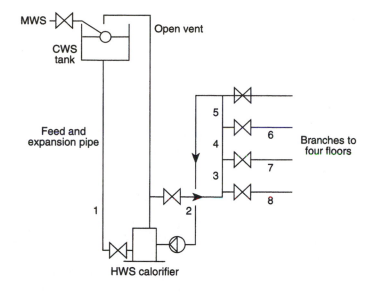

Figure 7.5   Centralized HWS system adopting the scale of demand units.

Table 7.10   Case Study 7.7

Section	1	2	3	4	5	6	7	8
Demand units	976	976	732	488	244	244	244	244
Simultaneous flow (litres/s)	2.6	2.6	2.1	1.5	0.9	0.9	0.9	0.9
Total flow (litres/s)	13.2	13.2	9.9	6.6	3.3	3.3	3.3	3.3

CONCLUSION

Clearly, the exponential effect, that the application of the usage ratio and the scale of demand units have, is considerable on simultaneous flow, on which pipe sizing is based. It is therefore important to apply common sense to this aspect of design. On the one hand, it is likely to be quite uneconomic as well as inappropriate to size the system using total flow rates. On the other hand, application of the scale of demand units reduces significantly the simultaneous flow in each pipe section. The scale of demand units may or may not be appropriate for the project in hand, and experience should be sought and, if necessary, applied if it differs widely from this methodology.

It was identified earlier that there is at least one exception in system design where total flow must be taken for sizing purposes. This occurs where groups of showers are present in, say, a sports centre or sports club or school. In these applications all the showers are likely to be in use simultaneously *and* continuously for more than one group of consumers.

You are also reminded of the range of usage ratios (P) given in the *CIBSE Guide* for water demand from 0.014 to 0.448.

7.4   Further reading

1.   CIBSE Guide Section B4
2.   CIBSE Guide G: Public health engineering

7.5   Chapter closure

You can now apply four methodologies for determining simultaneous flow in pipe networks. You are able to reduce a system having multiple values of probability P to a common usage ratio so that simultaneous flow can be determined in each pipe section. You are able to make decisions or, with the benefit of this underlying knowledge, seek advice from an experienced engineer relating to simultaneous flow in hot and cold water systems for various applications for the purposes of pipe sizing.

Hot and cold water supply systems utilizing the static head **8**

8.1   Introduction

Traditionally in the UK water supplies provided within the building originate from a storage tank located at high level, mains water being supplied to the tank from the rising main via a float-operated valve and drinking water being provided off the rising main at locations specified by the water undertaking. This gave the water undertakings some control over the number of connections off their mains water distribution network and hence limited the possibility of back-contamination of the water supply. In current design this is a good starting point, although there are variations around the country, and variations are required at any rate for high-rise buildings that exceed the static lift provided by the minimum pressure in the mains water supply at ground level.

For most buildings, some of the static head provided by the water storage tank located at high level in the building can conveniently be absorbed in sizing the pipework. In centralized hot water supply design, however, the weakest circuit should have available not less than a static head equivalent to 200 Pa/m. Anything below this can lead to uneconomically large pipes, and in such circumstances the use of a pump to overcome the hydraulic resistance in the index run should be considered. This is investigated in Chapter 9.

Sizing the plant for centralized hot water supply requires careful consideration of the following factors: recovery time, capital cost, operating costs, likelihood of service failure owing to abnormal loading and implications of siting plant in relation to the draw off points. The *CIBSE Guide G* sets out *plant sizing curves* under two headings: procedure for *constant tariff* fuel systems like gas and oil, and fuel tariff systems incorporating *off-peak periods* as in electric heating. Table 8.5, translated from the plant sizing curves, lists the hot water storage and boiler power per person based on a 2 h recovery period. Case Study 8.4 addresses the application of the constant tariff fuel system. You should note the discrepancies in hot water storage per person between Tables 8.2 and 8.5 since this affects the size of hot water storage plant and boiler plant.

8.2   Factors in hot water supply design

The following factors need consideration prior to proceeding with system design:

1. the number and type of fittings served (for example, for personal ablutions, for laundering, cooking, dishwashing);
2. number of consumers served;
3. simultaneous flow rates, which may require the application of probability or usage ratio P (see Chapter 7);
4. whether fittings are closely grouped or widely distributed (this will define the types of system as centralized hot water generation or point-of-use hot water);
5. nature of the water supply (this will define acceptable materials for the system and the potential need for water treatment);
6. storage temperature (this is now normally taken as 65 °C to inhibit the growth of *Legionella* spores).

Excessive storage temperatures increase the effects of scale formation and corrosion. The first three factors listed apply in addition to cold water supply. Factors 4, 5 and 6 apply only to hot water supply.

8.3   Design procedures

As implied in the introduction to this chapter, there are two methodologies for centralized systems: sizing on static head, and sizing on a pressure drop of around 300 Pa/m for pumped circulation. This chapter is confined to the former methodology. The design procedure is as follows:

1.   Adopt/select a routine for determining simultaneous demand, and calculate flow rates in all sections of the system (see Chapter 7).
2.   Identify the number of circuits in the system, and determine the index run from ratios of height to length (h/L) for each circuit in the system. The index circuit is that circuit whose ratio of h/L is numerically the lowest.
3.   Rank the remaining circuits from the next lowest value of h/L to the highest. They must then be sized in that order after the index run.
4.   Determine the available index pressure loss per metre in Pa/m from the available static head after making due allowance for the pressure required at the index fitting, and size the secondary outflow pipes of the index circuit, making sure not to exceed the available static head h.
5.   Assess the pressure available at the branches.
6.   Size the secondary outflow pipes in the remaining circuits in rank order.
7.   Size the secondary return and pump.
8.   Size the hot water storage calorifier.
9.   Determine the required net boiler power.
10.   Size the cold water storage tank.

Procedures 7, 8 and 9 do not apply to cold water supply; otherwise, the procedure is similar for both cold water supply and centralized hot water supply.

Case Study 8.1

Figure 8.1 shows a centralized system of hot water supply for a factory operating three shifts, which is occupied by 200 employees per shift. The system requires designing given the following information:

Fittings: 38 wash hand basins, 4 sinks and 14 showers.

Pipe section	1	2	3	4	5	6	7	8	9	10
Pipe length (m)	5	6	4	4	4	3	4	4	4	4

SOLUTION

The solution takes account of the factors regarding the choice of system listed at the beginning of this chapter and the design procedure.

Without any details of the nature of the water supply, we shall use copper tube Table X and allow 20% on straight lengths for losses

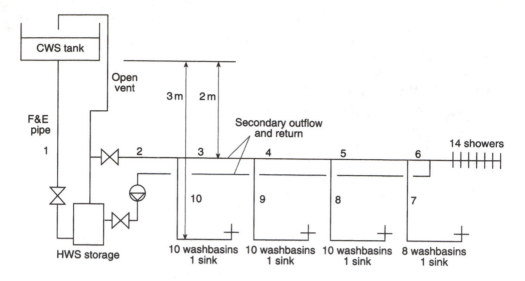

Figure 8.1   Case Study 8.1: centralized HWS system for a factory.

through fittings. The question of demand on the draw-off points must now be addressed. It is likely that all the showers will be in use together for each occasion of use. It is unlikely that all the wash handbasins will be in use simultaneously (although this point can be argued), and it is proposed that scales of demand units are applied to them and to the sinks (see Chapter 7). *CIBSE Guide G* or *CIBSE Concise Handbook table G3.2* gives data relating to the number of sanitary appliances allocated to staff in offices and shops, and this is related to the levels of occupancy and occupation.

The factory has a total of 38 wash basins, which accounts for the fact that we are dealing with an industrial building. Clearly a knowledge of the tasks being performed in the factory would be of benefit here.

It is proposed that the wash basins are taken on the congested scale of demand and the sinks, assumed here for cleaners' use, taken on the scale of public demand (see Table 7.8). Remember that assumptions should be kept to a minimum, with data from the water undertaking, architect or client used in preference.

DETERMINATION OF SIMULTANEOUS FLOW RATES

There are 14 showers discharging simultaneously at 0.15 litres/s each, giving a total flow of 2.1 litres/s. There are three branches of 10 wash basins and one sink, each giving a total of 122 DU, and one branch of eight wash basins and one sink giving 102 DU. This information is best tabulated (Table 8.1).

Table 8.1 Solution for Case Study 8.1

Section	1	2	3	4	5	6	7	8	9	10
Shower flow (litres/s)	2.1	2.1	2.1	2.1	2.1	2.1	–	–	–	–
Demand units	468	468	346	224	102	–	102	122	122	122
DU flow (litres/s)	1.4	1.4	1.2	0.8	0.5	–	0.5	0.5	0.5	0.5
Total simultaneous flow	**3.5**	**3.5**	**3.3**	**2.9**	**2.6**	**2.1**	**0.5**	**0.5**	**0.5**	**0.5**
Available pd (Pa/m)	308	308	308	308	308	308	4564	4815	5121	5255
Actual pd (Pa/m)	188	150	134	306	251	171	1175	1175	1175	1175
Flow diameter (mm)	**67**	**67**	**67**	**54**	**54**	**54**	**22**	**22**	**22**	**22**
Return diameter (mm)	–	35	35	28	22	–	–	–	–	–
TEL (m) (+20%)	6	7.2	4.8	4.8	4.8	3.6	4.8	4.8	4.8	4.8
pd (Pa)	1128	1080	643	1469	1205	616	–	–	–	–

NUMBER OF CIRCUITS AND THE INDEX RUN

There are five circuits in the system, and the h/L ratios are:

circuit 1,2,3,4,5 and 6	$h/L = 2/26 = 0.077$
circuit 1,2,3,4,5 and 7	$h/L = 3/27 = 0.111$
circuit 1,2,3,4 and 8	$h/L = 3/23 = 0.130$
circuit 1,2,3 and 9	$h/L = 3/19 = 0.158$
circuit 1,2 and 10	$h/L = 3/15 = 0.20$

RANKING THE CIRCUITS

The ratios are listed in rank order of magnitude, with the lowest numerically being the index circuit, namely 1,2,3,4,5 and 6. This circuit must be sized first, with the remaining circuits being sized in rank order from the next lowest h/L ratio progressively to the circuit with the highest numerical ratio. In fact it is not necessary to follow this ranking procedure here beyond the index circuit, owing to the system layout, but it would be wise nevertheless to keep to the ranking procedure.

INDEX PRESSURE AND INDEX PIPE SIZING

The available static head to the index circuit from Figure 8.1 is 2 m. Notice that it is taken from the underside of the CWS tank so that the system operates when the tank is almost empty of water. The static pressure is obtained from $P = h\rho g = 2 \times 1000 \times 9.81$. Thus available static pressure is 19 620 Pa. The minimum pressure at the index shower head is 10 kPa, so available static pressure for pipe sizing is reduced to 9620 Pa.

$$\text{Pressure loss per metre, pd} = \frac{P}{\text{TEL}} = \frac{9620}{26 \times 1.2} = 308 \, \text{Pa/m}$$

This is the average rate of pressure drop, which can be used to size the index circuit. Note that it is in excess of the minimum quoted in the introduction of 200 Pa/m. Clearly the available static pressure of 19 620 Pa must not be exceeded over the index run as there is no more static pressure available to use.

The index run can now be sized using Table X copper tube. This is shown in Table 8.1. The total pressure absorbed by the index run can now be calculated and compared with the static pressure available:

$$\text{Pressure absorbed} = (\text{pd in pipes 1 to 6}) + (\text{pd at index shower})$$
$$= 6141 + 10\,000 = 16\,141 \, \text{Pa}$$

$$\text{Static pressure available} = 19\,620 \, \text{Pa}$$

It is important not to absorb all the pressure available, as scale build-up will increase the pressure absorbed, and no account has been taken for the pressure drop across the secondary side of the storage vessel.

ASSESS THE PRESSURE AVAILABLE AT THE BRANCHES

The second circuit in rank order is 1,2,3,4,5 and 7. The available static head is 3 m and corresponding static pressure is 29 430 Pa. Pipe sections 1, 2, 3, 4 and 5 are part of the index run and already sized. The total pressure loss sustained in these pipes is 5525 Pa. The pressure available at branch 5/7 is therefore 29 430 − 5525 = 23 905 Pa.

The third circuit in rank order is 1,2,3,4 and 8. With a similar static head available, the static pressure is again 29 430 Pa. The pressure loss sustained in pipes 1,2,3 and 4, which are already sized, is 4320 Pa. The pressure available at branch 4/8 is therefore 29 430 − 4320 = 25 110 Pa.

The fourth circuit in rank order is 1,2,3 and 9 having an available static pressure of 29 430 Pa. The pressure loss sustained in pipes 1,2 and 3, which are already sized, is 2851 Pa. The pressure available at branch 3/9 is therefore 26 579 Pa.

The fifth circuit in rank order is 1,2 and 10 having an available static pressure of 29 430 Pa. The pressure loss sustained in pipes 1 and 2, which are already sized, is 2208 Pa. The pressure available at branch 2/10 is therefore 27 222 Pa.

SIZE THE REMAINING CIRCUITS

You will have noted that the only secondary outflow pipes left to size are branch pipes 7,8,9 and 10. We now have the pressures available at the branches for sizing these pipes. However, the pressure required at the index fitting in each branch should be accounted for before converting these available pressures to pressure losses per metre for pipe-sizing purposes. In each case it will be a sink tap and, like the basin tap, requires 2 kPa discharge pressure.

Thus:

$$\text{for pipe } 7, \text{pd} = \frac{23\,905 - 2000}{4 \times 1.2} = 4564\,\text{Pa}$$

$$\text{for pipe } 8, \text{pd} = \frac{25\,110 - 2000}{4 \times 1.2} = 4815\,\text{Pa}$$

$$\text{for pipe } 9, \text{pd} = \frac{26\,579 - 2000}{4 \times 1.2} = 5121\,\text{Pa}$$

$$\text{for pipe } 10, \text{pd} = \frac{27\,222 - 2000}{4 \times 1.2} = 5255\,\text{Pa}$$

These available pressures are shown in Table 8.1. Note how much higher in value they are than the available index pressure of 308 Pa/m. This is the result of the increase in static head from 2 m available to the index circuit to 3 m, for the remaining circuits. The remaining pipe sections can now be sized, and these are tabulated along with the actual rates of pressure loss.

Note that for a pipe size of 22 mm and a simultaneous flow of 0.5 litres/s, water velocity from the pipe-sizing tables is in excess of 1.5 m/s, which for copper pipe may generate noise. However, this must be put into context, as unlike a space heating system any such noise will be intermittent and therefore more acceptable than if it was continuous. Secondly, the system is installed in a factory in which there will be a degree of noise generation in any event.

As a general rule, limits on water velocity in hot and cold water service systems are not imposed owing to their intermittent use, when simultaneous flow is likely to occur only momentarily.

Excessive discharge pressure from hot water taps is potentially dangerous. There is an evidence for this in pipe sections 7,8,9 and 10, where only a fraction of the available pressure is absorbed by the branch pipes. As a general rule, the maximum pressure at a hot tap outlet should not exceed 50 kPa. In multi-storey buildings this limiting pressure is inevitably exceeded, and a pressure-limiting valve should be located behind the fitting.

SIZE THE SECONDARY PUMP

The secondary return is required to ensure hot water at or near the point of use at all times during occupation of the building. This is a regulatory requirement of the water undertaking to reduce the unnecessary consumption of water and to avoid a user waiting for hot water to discharge from the tap.

The secondary pump is required to circulate the water, as it is not usually possible to rely on natural circulation. Note that the pump is only used for this purpose when the system pipework is sized on static head.

It follows, therefore, that only sufficient water need to circulate to offset the heat loss from the circulating pipework. An alternative approach is to dispense with the secondary return and pump and fit electrical tracing tape to the secondary outflow pipe to maintain the supply temperature during periods when water is not being drawn off.

The heat loss from the secondary circulating pipework is a function of its sizes and extent and of the level of thermal insulation applied. It can be estimated using tables of heat loss from insulated pipe given in Section C3 of the *CIBSE Guide*. For this system it is estimated as 3 kW, and taking a temperature drop of 10 K across the circuit, the required mass flow to offset the heat losses is 0.071 kg/s.

Adopting a pressure loss of 250 Pa/m, the required size of the secondary return is 15 mm on 240 Pa/m using Table X copper tube. However, the use of 15 mm pipe for this purpose is not recommended owing to the probable effects of scale formation from the constant use of raw water. The minimum pipe size here should be 22 mm. In practice, however, the secondary return is usually taken as one to three sizes below that of the secondary outflow, and this is what is tabulated.

The pressure development required of the pump is based upon the loss sustained in 15 mm pipe. This will generously account for scale build-up between descaling maintenance. The length of the secondary return will be equivalent to sections 2, 3, 4 and 5, namely 18 m, and allowing for fittings the pump pressure required will be

$$P = \text{TEL} \times \text{pd} = 18 \times 1.2 \times 240 = 5184 \, \text{Pa}$$

Pressure loss through the secondary outflow, which forms part of the circulating pipework, is negligible as it has been sized on the simultaneous flow rates and not on 0.071 kg/s. Check this out using the pipe-sizing tables to confirm this point. Some allowance for the pressure drop across the secondary side of the hot water storage vessel should be made. This would be obtainable from the cylinder manufacturer. Here the required pump pressure for this reason will be increased from 5184 Pa to 7500 Pa.

Net pump duty: 0.071 kg/s at 7.5 kPa

On referring to pump manufacturers' literature you will find that it is often the smallest in the range. You will need to specify that it is to be used on secondary hot water, as manufacturing materials will be different from pumps used on closed systems.

SIZE THE HOT WATER STORAGE VESSEL

There are various ways in which this may be calculated. *CIBSE GUIDE G* gives measured daily hot water consumption in various types of building [*CIBSE Concise Handbook table G2.10* or *CIBSE Guide G*] and also in Table 8.2. Taking the figure of 15 litres/person for the factory with a total of 200 occupants per shift, the size of the hot water storage will be:

$$200 \times 15 = 3000 \, litres$$

This could be accommodated for practical considerations using two 1500 litres storage vessels:

Note: 15 litres/person means that the length of time for hot water discharge from a basin and shower is $15/0.15 = 100 \, s/person$ where 0.15 litres/s is the recommended discharge from a shower nozzle. On the other hand it is unlikely that all 200 workers will require a shower. You can see the importance of finding out how many of the 200 workers are doing activities which will require the use of the shower facilities.

NET BOILER POWER

This should be analysed by estimating the length of the recovery period required. For a three shift occupancy of eight hours a shift, a recovery period of two hours should ensure that hot water is available two hours

Table 8.2 Daily hot water demand

Building type	Hot water storage (litres/person/day)	Recovery period (h)
Boarding school	23	2.0
Day school	4.5	2.0
Dwellings	45	2.0
Factories (no canteen)	5/15	2.0
Hospitals	23/45	1.0/1.5
Hotels	36/45	1.0/1.5
Offices (no canteen)	4.5	2.0
Sports centres	40	1.0

into a shift. This is also the recommendation in [*CIBSE Concise Handbook table G2.10*, or *CIBSE Guide G*]

$$\text{Net boiler power} = \frac{\text{mass} \times \text{specific heat capacity} \times \text{temperature rise}}{\text{recovery time in seconds}}$$

$$= \frac{3000 \times 4.2(65 - 10)}{2 \times 3600}$$

$$= 96.25\,\text{kW}$$

For a single shift or two-shift operation the recovery period could be extended, thus reducing the required net boiler power.

SIZE THE COLD WATER STORAGE TANK

Provision of domestic storage to cover 24 h interruption of supply is given in the *CIBSE Guide G*, or *CIBSE Concise Handbook tables G2.2 and 3* and in Table 8.3. The water undertaking will specify for an identified locality the length of the interruption period so that the storage provision can be determined for a building using the data in the table.

There are 600 occupants per 24 hours in the factory. This may have a bearing upon the decision made by the water undertaking relating to the length of the interruption period. *CIBSE Concise Handbook tables G2.2 and 3*, or *CIBSE Guide G* does not give a figure for factories for fairly obvious reasons as it will depend upon the nature of the activities. Table 8.3 gives a figure of 15 litres/person. The Occupier should be able to identify the number of workers needing the use of the shower facilities. We shall assume a 24 h storage for hot and cold water of 50 litres/person.

Table 8.3   Domestic cold water storage for a 24 h interruption

Type of building	Storage (litres)
Dwellings up to 4 bedrooms	120/bedroom
Dwellings over 4 bedrooms	100/bedroom
Hotels	135/bedroom
Offices with canteens	45/person
Offices without canteens	40/person
Restaurants	7/meal
Schools	
Boarding	90/person
Day	20/person
Factories	15/person

We shall also assume here that an 8 h interruption is required by the water undertaking, in which case the cold water storage requirement will be: $600 \times 50 \times 8/24 = 10\,000$ litres.

CONCLUSION

You will notice that part of the design solution requires the application of some common sense: for example, the effect of potential scale build-up on pipe sizes and how much of the available static head can be absorbed in sizing the secondary outflow. It is important to stress the need for these attributes rather than blindly adopting a recommended design procedure. This is particularly true when considering how to arrive at simultaneous flow rates. Adopting a design that has success-fully worked on an earlier similar project is not an admission of failure!

The following is an example of the application of common sense.

Case Study 8.2

A group of 22 showers, each rated at 0.15 litres/s, is attached to a school gymnasium whose timetable has four groups of students using them equally spaced over a 6 h day. The ratio of static head to index length is 0.09, and the cold water storage tank must be sized on an 8 h inter-ruption of mains water service. If the storage per student is 15 litres, estimate without the use of tables the size of: (a) the calorifier, (b) the boiler, (c) the storage tank and (d) the cold feed and secondary outflow.

SOLUTION

(a)   If four groups of students use the showers over a 6 h period, the calorifier may be sized to supply each group, in which case the recovery period must be limited to

$$\frac{6}{4} = 1.5\,\text{h}$$

Assuming all the showers could be in use on each occasion:

$$\text{net size of the calorifier} = 22 \times 15 = 330\,\text{litres}$$

(b)   For a 1.5 h recovery:

$$\text{net boiler power} = \frac{330 \times 4.2(65 - 10)}{1.5 \times 3600} = 14.1\,\text{kW}$$

(c)   An 8 h interruption, at worst, would occur at the beginning of the school day, so the storage tank must be sized to cater for a day's use of the showers:

$$\text{Tank size} = 4 \times 22 \times 15 = 1320 \, \text{litres}$$

(d)   The simultaneous flow will be equivalent to all the showers discharging together:

$$22 \times 0.15 = 3.3 \, \text{litres/s}$$

The available static pressure for sizing the cold feed and the secondary outflow will be

$$h/L \times \rho \times g = 0.09 \times 1000 \times 9.81 = 883 \, \text{Pa/m}$$

Note: The ratio h/L is a static height available to index length.

Using Box's formula, which is an adaption of the D'Arcy equation for turbulent flow in pipes [*Heat and Mass Transfer in Building Services Design*]

$$\frac{dp}{L} = \frac{(\text{VFR})^2 f \rho g}{3d^5}$$

where the frictional coefficient f may be taken as 0.005 and density of water at 65 °C as 980 kg/m³. Then, rearranging the formula in terms of diameter d:

$$d = \left(\frac{(\text{VFR})^2 f \rho g}{3 dp/L} \right)^{1/5}$$

and substituting,

$$d = \left(\frac{(0.0033)^2 \times 0.005 \times 980 \times 9.81}{3 \times 883} \right)^{1/5}$$

from which

$$d = 0.0456 \, \text{m} \quad \text{or} \quad 46 \, \text{mm}$$

The nearest standard diameter in copper to Table X is 54 mm.

Referring now to *CIBSE* tables for Table X copper pipe: for a flow rate of 3.3 litres/s on 54 mm diameter, pressure drop is 386 Pa/m at 75 °C for the secondary outflow and 483 Pa/m at 10 °C for the cold feed. You should now confirm these pipe sizes.

Note: There are no pipe-sizing tables or corrections for water at 65 °C.

BOILER POWER ASSOCIATED WITH THE GENERATION OF HOT WATER

If the hot water supply system is interconnected with the space heating system the net boiler power BP for the generation of hot water is dependent upon the length of the regeneration period:

$$\text{Net BP} = \frac{m \times C \times dt}{\text{time in seconds}} \quad (\text{kW})$$

This calculation is done for both Case Studies 8.1 and 8.2. If primary water is used in the heat exchanger within the calorifier, its mean temperature should be at least 10 K above that of the secondary storage temperature of 65 °C for good heat transfer from the primary medium to the secondary medium. It is sensible to have a summer boiler sized to cope with the hot water supply so that the heating boiler plant is not operating during the summer months. Modular boiler plant can be arranged so that one or more of the boiler modules will match the primary hot water demand efficiently. The mass flow of primary water, M, to transfer the heat energy from the boiler to the primary heat exchanger in the calorifier is calculated from:

$$M = \frac{\text{net BP } plus \text{ margin}}{4.2(t_f - t_r)}$$

where t_f and t_r are the primary flow and return temperatures to the heat exchanger.

Case Study 8.3

Figure 8.2 shows a centralized HWS system in elevation serving one-bedroom nurses' accommodation for 12 occupants. In accordance with the design procedure set out in this chapter, size the system. The local water undertaking requires a 12 h interruption of supply to be accounted for when sizing the storage tank. Allow 20% on straight pipe for fittings. Copper tube to Table X shall be used. The index draw-off to each student's accommodation is a shower fitting, which requires an operating pressure of 10 kPa.

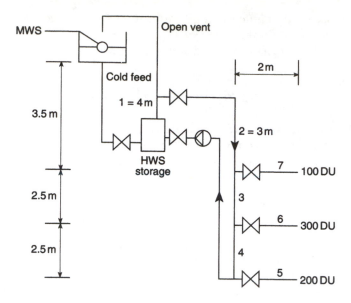

Figure 8.2 Case Study 8.3: sizing centralized hot water supply to student accommodation.

SOLUTION

The solution relating to pipe sizes is given in Table 8.4. You should now work through your own solution and see if you agree. From *CIBSE Concise Handbook table G2.10*, or *CIBSE Guide G* or Table 8.2, calorifier size $= 12 \times 45 = 540$ litres

$$\text{Net boiler power} = \frac{540 \times 4.2(65 - 10)}{2 \times 3600} = 17.3\,\text{kW}$$

Table 8.4 Case Study 8.3: solution

Section	1	2	3	4	5	6	7
Demand units	600	600	500	200	200	300	100
Simultaneous flow (litres/s)	1.7	1.7	1.5	0.8	0.8	1.0	0.5
Available pd (Pa/m)	2253	2253	7155	11224	11224	7155	2253
Secondary flow dia (mm)	**35**	**35**	**35**	**22**	**22**	**28**	**22**
Actual pd (Pa/m)	1330	1067	850	Off table	Off table	1200	1175
TEL (+20%) (m)	4.8	3.6	3	3	2.4	2.4	2.4
Pressure drop (Pa)	6384	3841	2550				
Secondary return diameter (mm)	–	**22**	**22**	**22**	–	–	–

Net pump duty:

$$M = \frac{\text{estimated 2 kW circulation heat loss}}{4.2 \times 10} = 0.048\,\text{kg/s}$$

$P = (124\,\text{Pa/m on 15 mm tube at 0.048 kg/s}) \times 8 \times 1.2 = 1190\,\text{Pa}$
Allowing for the pressure drop on the secondary side of the calorifier

$$P = 3.5\,\text{kPa}$$

Net pump duty is 0.048 kg/s at 3.5 kPa. Cold water storage from
CIBSE Concise Handbook tables G2.2 and 3, or *CIBSE Guide G* or
Table 8.3, nurses' homes/dwellings have 120 litres/bed.

$$\text{CWS tank size} = 12 \times 120 \times \frac{12}{24} = 720\,\text{litres}$$

8.4   Plant sizing guides

CIBSE Guide G includes plant sizing guides for different building types
including schools, hotels, restaurants, offices, shops and student hostels.
The sizing routine can be done for hot water storage vessels having upper
and lower electric heating elements and for plant operating on natural gas or
fuel oil. The routine estimates the size of the hot water storage and the
output of the heater. Service and catering curves allow separate calculation.
The *CIBSE Guide* suggests an allowance of 25% in sizing the hot water
storage to allow for the mixing of incoming cold water if the storage vessel
does not incorporate a device to inhibit mixing.

Case Study 8.4

An office has an occupancy of 140 staff and the centralized hot water
supply to the wash hand basins shall be at 55 °C with a 2 h recovery.
Estimate the size of the hot water storage and the boiler rating adopting the appropriate plant sizing curves.

SOLUTION

From the curve for offices [*CIBSE Concise Handbook tables G2.16
(a) and (b) or CIBSE Guide G*] or Table 8.5, hot water per person =
1.2 litres and net boiler power at a storage temperature of 65 °C per
person = 0.04 kW.

Net boiler power correction for a storage at 55 °C $= 0.04\left(\dfrac{55 - 10}{65 - 10}\right)$

$$= 0.0327\,\text{kW/person}$$

Net boiler rating $= 0.0327 \times 140 = 4.58\,\text{kW}$. If heat losses are esti-mated at $0.5\,\text{kW}$ the final boiler rating will be $5.08\,\text{kW}$. Storage capacity $= 1.2 \times 140 = 168\,\text{litres}$ and if 25% is added to allow for mixing final storage capacity $= 168 \times 1.25 = 210\,\text{litres}$.
Note:

1. The *service* requirement for offices has been calculated here. If the office has catering facilities the storage and boiler power require-ments for catering need calculating and *adding* to the service requirements (see Table 8.5).
2. If the above solutions are now compared with the data in Table 8.2 and *CIBSE Concise Handbook table G2.10*, or *CIBSE Guide G* for offices:

$$\text{Storage capacity} = 140 \times 4.5 = 630\,\text{litres}$$

and

$$\text{boiler power} = \frac{630 \times 4.2(55 - 10)}{2 \times 3600} = 16.5\,\text{kW}$$

Clearly there is a serious discrepancy here between the two sizing routines with the current plant sizing curves giving much lower storage and boiler power. Table 8.5 is based on the plant sizing curves in *CIBSE Guide G* for a 2 h recovery period. The factor that has a direct

Table 8.5 Hot water storage and boiler power based on *CIBSE Guide G* plant sizing curves for a two hour recovery and 65 °C storage temperature

Type of building	Storage capacity (litres/person)	Boiler output (kW/person)
Schools		
Service	1.15	0.03
Catering	2.3	0.07
Hotels		
Service	15	0.50
Catering	4.6	0.14
Restaurants		
Service	0.52	0.03
Catering	0.85	0.03
Offices		
Service	1.20	0.04
Catering	2.40	0.13
Shops		
Service	1.90	0.10
Catering	1.60	0.07
Student hostels		
Service	7.00	0.23

impact on the results is the amount of hot water allowed per person namely 1.2 litres or 4.5 litres. It would appear from the historical evidence that use of hot water in offices and the other building types listed in Table 8.5 was over-estimated in the past but this should not deter you from a common sense approach.

8.5   Direct fired and point of use heaters

GAS-FIRED WATER HEATERS

This method of independently heating domestic hot water directly instead of indirectly via a water-to-water or steam-to-water heat exchanger in the calorifier is now frequently specified. Current policy favours the separation of hot water generation and space heating on the grounds of increased thermal efficiency. It also allows the location of the gas-fired storage heaters at or nearer the points of use rather than in the boiler plant room.

Manufacturers offer proven high thermal and combustion efficiencies for their products and supply the heaters for immediate connection to the rising main and gas supply either piped or bottled. Design simultaneous flow is achieved by connecting the appropriate number of heaters in banks to common outflow and return headers (see Figure 6.13).

POINT-OF-USE HOT WATER HEATERS

If fittings are widely distributed it may not be viable to have long runs of secondary outflow and return pipework serving them. In such cases the use of local hot water heaters is justifiable. These may be gas fired or electrically operated, and they come in different forms of which there are four main types:

- *instantaneous*: suitable for hand washing at one draw-off point and connected directly to the rising main;
- *over/under sink*: storage heaters ranging from 5 to 30 litres serving one draw-off point and connected directly to the rising main;
- *multipoint*: storage heaters for serving a group of fittings and connected to a high-level storage tank;
- *cistern multipoint*: storage heaters with integral supply tank for serving a group of fittings and connected to the rising main.

Manufacturers are keen to recommend type, size and duty for specified applications.

8.6   Chapter closure

You are now able to design hot and cold water supply systems utilizing the static head imposed by the cold water storage tank for pipe sizing. You are able to size the cold water storage tank, hot water storage vessel, net boiler power and secondary pump, which offsets the heat loss in the circulating pipework. In conjunction with Chapter 7 you are able to assess the likely simultaneous flow in the secondary circuit for the purposes of pipe sizing and in a multi-circuit system put the circuits in rank order for pipe-sizing purposes.

Remember the dictum of applying common sense to the determination of simultaneous flows, hot water storage and recovery time and hence boiler power for centralized systems. You can apply the knowledge you have now gained to sizing point-of-use hot water heaters and associated pipework. You are advised, however, to take account of the heater manufacturer's recommendations.

Hot and cold water supply systems using booster pumps 9

Nomenclature

CWS	cold water service
dp	pressure drop (Pa)
DU	demand units
E	drinking water volume (litres, m^3)
F&E	feed and expansion
g	gravitational acceleration (9.81 m/s^2)
h	static head (m)
HWS	hot water service
M	mass flow rate (kg/s)
MWS	mains water service
P	pressure (Pa)
pd	pressure drop (Pa, Pa/m)
u	velocity (m/s)
V	volume (m^3)
VFR	volume flow rate (m^3/s)
ρ	density (kg/m^3)
$\sum$	sum of

9.1 Introduction

For centralized hot water, the economic minimum *available* pressure drop for the index circuit and therefore all the other circuits in the system is about 200 Pa/m. When selecting the pipe size the nearest diameter may yield a pressure drop lower than 200 Pa/m. If the static head is insufficient to provide this pressure for the index circuit, a pump should be considered for the secondary circuit to overcome the hydraulic resistance in the index run.

If there is no static head available to a centralized system of hot water supply, as would be the case in the storage of cold water on the same level as the calorifier, a pressurization unit (consisting of a booster pump, pressure vessel and pressure switches) and an expansion vessel located in the cold feed are required to provide the static lift, in addition to the secondary HWS pump. The expansion vessel is required to accept expansion water from the HWS calorifier during the recovery period.

In the case of the supply and storage of cold water for sanitary and other uses, when a building exceeds the static lift of the incoming mains supply pressure, a booster pump is required to get the water to the upper storeys that are beyond the reach of the mains water static lift, and to the high-level storage tank.

Sometimes part of the cold water storage capacity for a building must be held at ground or basement level, owing to the sheer mass of water requiring storage. The water undertaking may also make this stipulation. Clearly, in this case a pressurization unit consisting of a booster pump, pressure vessel and pressure switches is required to supply the water to all the upper storeys via the high-level tank and to account for the supply of drinking water beyond the mains water static lift. It is important to do some research of plant and equipment used in boosted water supply. This will add to your knowledge of this subject [*Building Services OPUS Design File*].

9.2   Pumped hot water supply

Case Study 9.1

Figure 9.1 shows a centralized hot water supply system serving a five-storey building with 3 m floor heights.

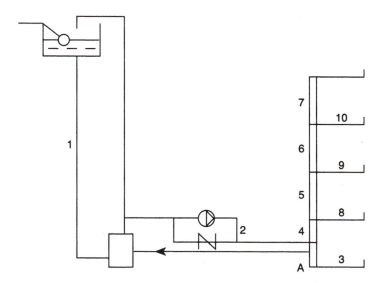

Figure 9.1   Case Study 9.1: centralized pumped hot water supply: A, regulating valve on secondary return. All other valves omitted.

(a)   Show that the system should not be sized on the available static head.

(b)   size the pipework and secondary pump. Fittings allowance on straight pipe is 25%. The index fittings are basin/sink taps and therefore have a discharge pressure of 2 kPa. Tubing shall be copper to Table X [*CIBSE Concise Handbook Section C*]. Each branch carries 100 demand units.

DATA RELATING TO FIGURE 9.1

Vertical height from highest draw-off point to underside of tank is 1 m.

Section	1	2	3	4	5	6	7	8	9	10
Pipe length (m)	15	10	5.5	1.5	3	3	7	4	4	4

SOLUTION

Note how the secondary pump is located in the outflow pipe. If the pump delivery is exceeded by the simultaneous flow it can flow through the pump bypass. As the pump is sized on the simultaneous flow this should happen only infrequently.

(a)   Following the procedure and case studies in Chapter 8, the index circuit consists of sections 1, 2, 4, 5, 6 and 7, and the pressure drop available from the static head of 1 m is approximately 152 Pa/m. Check this calculation for yourself. This is below the 200 Pa/m threshold, and therefore the system should be pumped.

It could be strongly argued, however, that as there is only one float here attracting a low index pressure it is not worth going to the expense of a pump. In practice there could be many floats at this level, and sizing for a pump would be justified. The solution will proceed on this basis so that you will know how to undertake the design in such an instance.

(b)   Adopting a pressure of 300 Pa/m the index circuit can be sized. The solution is tabulated in Table 9.1. Note however that Section 1 does not form part of the index circuit for the pump and must be sized using some of the available static pressure to the index circuit. The pressure loss is tabulated for the cold feed and expansion pipe and secondary flow only. The secondary return is estimated at one to two sizes below the outflow to maintain hot water at the points of use. The pressure drop in the secondary return is estimated as the same as that in the corresponding outflow pipe sections.

Table 9.1 Case Study 9.1

Section	1	2	3	4	5	6	7	8	9	10
Demand units	500	500	100	400	300	200	100	100	100	100
Simultaneous flow (litres/s)	1.5	1.5	0.5	1.3	1.0	0.8	0.5	0.5	0.5	0.5
Adopted (Pa/m)	152	300	874	300	300	300	300	1104	800	595
Secondary flow diameter (mm)	**54**	**42**	**28**	**42**	**35**	**35**	**28**	**28**	**28**	**28**
Actual pd (Pa/m)	120	332	340	257	406	273	340	340	340	340
TEL (flow only) (m)	18.8	12.5	6.88	1.9	3.75	3.75	8.75	5	5	5
pd (flow only) (Pa)	2250	4150		488	1523	1024	2975			
Secondary return diameter (mm)	–	**28**	**22**	**28**	**28**	**22**	**22**	–	–	–

Net pump duty

Pressure loss in the index outflow sections $= 10\,160\,\text{Pa}$

Estimated pressure loss in the secondary return $= 10\,160\,\text{Pa}$

Total pressure loss in the index pipework $= 20\,320\,\text{Pa}$

Pump pressure required $= 20\,320 + \text{pd at index fitting} + \text{pd in}$
hot water storage vessel

$= 20\,320 + 2000 + 3000 \ (\text{estimated})$

$= 25\,320\,\text{Pa}$

Net secondary pump duty $= 1.5\,\text{litres/s at } 26\,\text{kPa}$

The pressure drop on the secondary side of the hot water storage vessel consists of two shock losses: a sudden contraction ($0.5u^2/2g$) at the secondary outlet and a sudden enlargement ($u^2/2g$) at the secondary return inlet, assuming water velocity within the storage vessel approaches zero. If water velocity is $1.0\,\text{m/s}$ in each of the secondary outflow and return pipes, the pressure loss is approximately $735\,\text{Pa}$. Check this for yourself using [*Heat and Mass Transfer in Building Services Design*, Spon Press]. Clearly, $3000\,\text{Pa}$ is quite sufficient.

Note that the pump pressure does not include pressure to overcome the static lift from the pump outlet to the highest fitting. It is provided in this case study by the static head imposed on the system from the high-level storage tank.

If the hot water storage vessel was supplied from a cold water storage tank on the same level, a pressurization unit would be required to lift the water to the highest fitting (see Figure 9.2). If the hot water storage vessel is supplied direct off the rising main, minimum mains pressure must be sufficient to overcome the static lift, otherwise a pump is required.

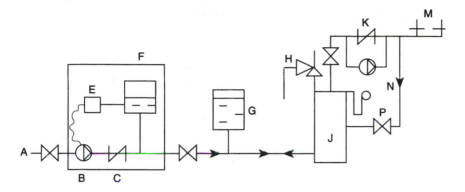

Figure 9.2 Centralized HWS system pressurized at the cold feed: A, low-pressure cold water supply; B, booster pump; C, non-return valve; E, pressure switch; F, pressure vessel; G, expansion vessel; H, pressure relief valve; J, HWS storage vessel; K, non-return valve; M, draw-off points; N, HWS pump; P, pressure gauge. Note: B, C, E and F are the constituent parts of the pressurization unit.

Sizing the F&E pipe, section 1

The cold feed is sized on some of the available static head. The actual rate of pressure loss in section 1 from Table 9.1 is 120 Pa/m. This is equivalent to

$$120 \times 15 \times 1.25 = 2250 \, \text{Pa}$$

and the corresponding static head absorbed will be

$$\frac{2250}{1000 \times 9.81} = 0.23 \, \text{m}$$

Sizing the secondary flow, branch 3

Flow in branch 3 and float at junction 4/3 is now considered, taking into account hydraulic balancing with the index run:

$$\text{Pump pressure available} = \text{that absorbed in sections}$$
$$4, 5, 6 \text{ and } 7 \text{ (flow only)}$$
$$= 488 + 1523 + 1024 + 2975$$
$$= 6010 \, \text{Pa}$$

$$\text{Pressure available to size flow branch } 3 = 6010/6.88$$
$$= 874 \, \text{Pa/m}$$

and the nearest pipe size from Table X is 28 mm at 340 Pa/m.

Sizing the secondary flow, branch to float 8

Flow in float 8 at junction 4/5 is considered next

$$\begin{aligned}
\text{Pump pressure available} &= \text{that absorbed in sections} \\
&\quad \text{5, 6 and 7 (flow only)} \\
&= 1523 + 1024 + 2975 \\
&= 5522\,\text{Pa}
\end{aligned}$$

$$\text{Pressure available to size float } 8 = \frac{5522}{5} = 1104\,\text{Pa/m}$$

and the nearest pipe size from Table X is 28 mm at 340 Pa/m.

Sizing the secondary flow, branch to float 9

Secondary flow in float 9 at junction 5/6 is now considered:

$$\begin{aligned}
\text{Pump pressure available} &= \text{that absorbed in sections} \\
&\quad \text{6 and 7 (flow only)} \\
&= 1024 + 2975 \\
&= 3999\,\text{Pa}
\end{aligned}$$

$$\text{Pressure available to size float } 9 = \frac{3999}{5} = 800\,\text{Pa/m}$$

and the nearest pipe size is 28 mm at 340 Pa/m.

Sizing the secondary flow, branch to float 10

Pump pressure available at junction 6/7 is that absorbed in section $7 = 2975\,\text{Pa}$

Pump pressure available to size float $10 = 2975/5 = 595\,\text{Pa/m}$

and the nearest pipe size is 28 mm at 340 Pa/m.

Circuit balancing

You will see that the attempt to balance hydraulically in the choice of pipe size has not been successful and floats 3, 8, 9 and 10 will need regulation. For example, regulation on float 3 will be $(874 - 340)6.88 = 3674\,\text{Pa}$. It is left to you to determine the regulating pressures required on floats 8, 9 and 10.

Note: When draw-off points are in use, floats 3, 8, 9 and 10 are subject to the combined effect of pump *and static* pressure. This will have a significant influence upon the discharge pressure from the draw-off points. If floats 3, 8 and 9 are sized on the combined pressure of pump and static the pipes reduce in size from 28 to 22 mm.

Excessive draw-off pressures

When a draw-off point is in use, it is influenced by the algebraic sum of static and pump pressures. The effect in a multi-storey building having a centralized hot water supply system results in excessive pressures at the draw-off points on the lower floors. Pressure-limiting valves should be fitted when discharge pressures are likely to exceed 250 kPa at the draw-off point to avoid splash and, in the case of HWS, consequent scalding.

Use of the HWS ring main

In commercial applications such as laundries, hot water supply is often direct rather than indirect, and use is made of a ring main to supply the draw-off points.

Case Study 9.2

Figure 9.3 shows an HWS ring main system. Size the ring main and branches and size the pump. Assume that the static head from the

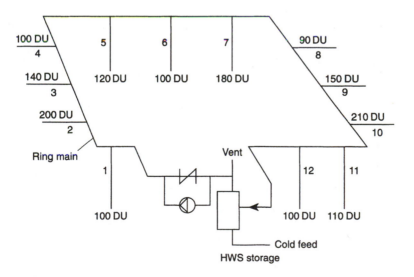

Figure 9.3   Case Study 9.2: HWS ring main system.

CWS tank is insufficient for sizing the pipework but can provide the static lift.

DATA

Total equivalent length of the ring main is 48 m.
Pressure required at each draw off point is 10 kPa.
Adopt a pressure drop of 300 Pa/m.

SOLUTION

Total demand units $\sum(DU) = 1600$, and from Table 7.9 simultaneous flow is 3.8 litres/s. From the pipe-sizing table for copper Table X [*CIBSE Concise Handbook Section C*], the ring main diameter selected is 54 mm on 500 Pa/m. The velocity is approximately 1.7 m/s, but flow of 3.8 litres/s is the *total* simultaneous flow for the ring main and immediately reduces as the hot water flows around the ring, thus reducing the pressure loss progressively back to the HWS storage vessel. Size of the ring main is 54 mm.

The branches can be sized using copper Table X after converting the demand units to simultaneous flow (Table 9.2).
Pump duty:

pressure developed = pd in index circuit *plus* pd at index terminal

Note: Pressure loss in the index branch is ignored, as it is assumed to be short. The pressure loss in the ring main of 500 Pa/m reduces progres-

Table 9.2   Case Study 9.2

Branch reference	DU	Flow (litres/s)	Diameter (mm)	pd (Pa/m)
1	100	0.5	28	340
2	200	0.8	35	273
3	140	0.6	35	163
4	100	0.5	28	340
5	120	0.5	28	340
6	100	0.5	28	340
7	180	0.7	35	215
8	90	0.5	28	340
9	150	0.6	35	163
10	210	0.8	35	273
11	110	0.5	28	340
12	100	0.5	28	340

sively around the ring as flow reduces. An average pd of 350 Pa/m will be used to assess the pump pressure.

$$\text{Pressure developed} = (350 \times 48 + 10\,000)\,\text{Pa}$$
$$= 26.8\,\text{kPa}$$
$$\text{Simultaneous flow} = 3.8\,\text{litres/s}$$

Thus net pump duty is 3.8 litres/s at 27 kPa.

Note: The pump flow must be oversized, not least as the flow must be sufficient to ensure some water returns to the storage vessel when the simultaneous flow of 3.8 litres/s is drawn off the ring main.

Case Study 9.3

Figure 9.4 shows an HWS ring main serving ten draw-off points. Size the ring main and branches. Size also the pump, calorifier and electric immersion heater. Assume the available static head from the CWS tank is insufficient for pipe sizing but can provide the static lift.

DATA

Total equivalent length of the ring main is 20 m.
Pressure required at the draw-off points is 15 kPa.
Number of people using the system simultaneously is 30.
Hot water storage is 25 litres/person.
Regeneration period is 1 h.
Flow rate to each draw-off point is 0.2 litres/s.

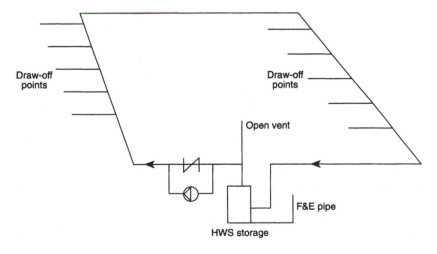

Figure 9.4   Case Study 9.3: HWS ring main serving 10 draw-off points.

SOLUTION

Clearly the design solution to hot water supply is made easier if all the parameters are known, as they are here.

Simultaneous flow in the ring main is 2.0 litres/s and using Table X copper tube [*CIBSE Concise Handbook Section C*] the diameter is 54 mm at a pressure loss of 156 Pa/m and a velocity of 1.0 m/s or 42 mm diameter pipe at a pressure loss of 560 Pa/m and a velocity of approximately 1.7 m/s. If all the fittings are used continuously as well as simultaneously there will be the possibility of persistent noise generation if 42 mm pipe is adopted. This may not be important if background noise is present. This solution will take a ring main size of 42 mm.

At 0.2 litres/s, branch size from Table X [*CIBSE Concise Handbook Section C*] will be 22 mm at a pressure loss of 228 Pa/m.

Pump pressure = index pressure loss plus pressure loss at index fitting, index branch ignored. Taking an average pd around the ring main of 400 Pa/m = 400 × 20 + 15 000 = 23 000 Pa.

$$\text{Flow rate} = 2.0 \text{ litres/s}$$

Net pump duty is 2.0 litres/s at 23 kPa. The comments referring to pump duty in the solution to Case Study 9.2 are the same here.

$$\text{Calorifier size} = 30 \times 25 = 750 \text{ litres}$$

$$\text{Net electrical power} = \frac{750 \times 4.2(65 - 10)}{3600} = 48.1 \text{ kW}$$

9.3   Boosted hot water supply

If the cold water supply comes from ground/basement storage there is little or no available static pressure, and the hot water supply system must therefore be pressurized. Consider the centralized system in Figure 9.2.

The system shown is appropriate where the cold water storage tank is located at the same level as the HWS storage vessel at the lowest point in the system or when the minimum MWS pressure is very low. A pressurization unit is located in the cold feed, and a booster pump handles cold water, operating intermittently under the dictates of a pressure switch. The integral pressure vessel has a gas cushion that may have a membrane to divide the gas from the water.

Net pressure developed by the booster pump must include the static lift to the highest fitting. Net booster pump flow rate is equivalent to the simultaneous flow of hot water supply and therefore cold water make-up.

In addition, a further pump is located in the secondary circuit. Its flow rate is the simultaneous flow for the system, and its developed pressure accounts for the sum of the hydraulic loss in the index circuit and the discharge pressure at the index draw-off point.

Note that in addition to the pressure vessel, which forms an integral part of the pressurization unit, there is also an expansion vessel fitted to the cold feed. This accepts expansion water from the HWS storage vessels during the recovery period. It can be sized from Table 4.1.

9.4   Boosted cold water supply

With the advent of high-rise buildings, it is essential to ensure that the mains water service is sufficiently pressurized to overcome the static height so as to supply water at a reasonable pressure to the highest draw-off point in the building, which might be the ball valve in a roof-level storage tank. If minimum MWS pressure at ground level is 2.5 bar gauge, this is equivalent to a static lift of 25.5 m. Given floor heights of 3 m, the main can supply seven storeys under its own pressure and still have some surplus pressure available for hydraulic resistance to flow and discharge.

If the building is over seven storeys the MWS must be boosted to the upper floors. The need for water, apart from industrial applications, which may require hot water for process as well, is twofold:

- for a potable water supply suitable for cooking and drinking;
- for a supply of water for washing and sanitary use.

A further requirement may include a water supply for fire hose reels and/or sprinkler system and/or wet fire riser.

There is therefore a need for at least two water supplies within the occupied building. For high-rise buildings that are beyond the static lift of the rising main, there may be a need therefore for the provision of booster pumps serving two or three independent services.

The ground storage of water is becoming the norm for high-rise buildings, and for domestic consumption (drinking, cooking, washing and sanitary use) a common rule of thumb is 2/3 ground storage and 1/3 roof level storage. Fire hose reel, wet fire riser and sprinkler systems may also be served from ground storage tanks. In each of these cases booster pumps are required.

For domestic water use in larger buildings beyond the static lift of the MWS there are two systems of water boosting in use: control by water level and control by water pressure.

BOOSTED WATER: CONTROL BY WATER LEVEL

A typical arrangement is shown in Figure 9.5. There is a tendency to flow reversal, owing to the effects of gravity in the boosted riser when the pump stops. The recoil valve (F) is a spring-loaded non-return valve, which is fitted to insure against this effect. Pipes K and P will require water pressure reduction at intervals of five storeys to avoid excessive pressures at the draw-off points. A fixed pressure-reducing valve is used for this purpose.

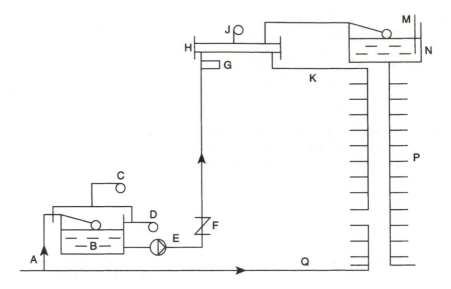

Figure 9.5 Boosted cold water, control by water level, valves omitted: A, mains water service; B, ground storage tank; C, open vent with filter; D, overflow with filter; E, booster pump; F, recoil valve; G, float switch; H, drinking water header; J, air release/intake valve; K, drinking water to upper floors; M, float switch; N, high-level cold water storage tank; P, cold water down-service serving urinals, WCs, basins, sinks etc.; Q, drinking water to lower floors.

Referring further to Figure 9.5, a branch from the rising main serves the drinking water points to four lower floors. Another branch from the rising main serves the ground storage tank via a ball valve. This tank in fact carries the drinking water to the upper floors as well as the water for other uses, and therefore must be protected from ingress of foreign matter. The booster pump is activated by either of the float switches: at low water level in the high-level storage tank, which serves all 13 floors with water for general use, or low water level in the drinking water header, which supplies the nine upper floors with potable water.

Case Study 9.4

Consider a 15-storey building with four 3-bedroom flats per floor, each flat having one drinking water point. Minimum mains water pressure is 2 bar gauge and floor heights are 3 m.
    Determine:

(a) cold water storage requirement;
(b) net booster pump duty;
(c) the drinking water storage.

SOLUTION

A typical riser diagram for the system is similar to Figure 9.5 except in relation to the number of storeys in the building.

(a)   From *CIBSE Guide [Concise Handbook Section G]* the recommended storage of cold water for a 24 h interruption of supply is 120 litres per bedroom.

$$\text{Gross storage} = 120 \times 3 \times 4 \times 15 = 21\,600 \text{ litres}$$

The water undertaking will advise the length of the interruption period, which we shall take here as 12 h. Thus gross storage is $21\,600 \times 12/24 = 10\,800$ litres.

Applying the rule of thumb:

$$\text{Ground storage} = 10\,800 \times \frac{2}{3} = 7200 \text{ litres}$$

$$\text{High level storage} = 10\,800 \times \frac{1}{3} = 3600 \text{ litres}$$

(b)   Net pump flow rate consists of the maximum drinking water demand *plus* the supply for refilling the high-level storage tank. The lower floors can be supplied from the rising main.

$$\text{Static lift of rising main} = \frac{P}{\rho g} = \frac{200\,000}{1000 \times 9.81} = 20.4 \text{ m}$$

This is equivalent to six storeys; thus nine storeys must be pumped. This represents $9 \times 4 = 36$ drinking water points, and taking 10 demand units per point we have 360 demand units.

From Chapter 7 the simultaneous flow for 360 DU is 1.2 litres/s. If the booster pump must refill the tank in 2 h the water supply will be

$$\frac{3600}{2 \times 3600} = 0.5 \text{ litres/s}$$

Thus the pump flow rate is

$$1.2 + 0.5 = 1.7 \text{ litres/s}$$

Note: If each flat has one bath, two basins, one sink and one shower, demand units total 120, and for the entire building total $DU = 120 \times 4 \times 15 = 7200$.

From Section *B4* of the *CIBSE Guide* this represents a simultaneous flow of 13.8 litres/s from the high-level tank. However, simultaneous flow is not continuous flow and will only occur momentarily, so it must not be confused with the calculated fraction (0.5 litres/s for 2 h) for replenishing the high-level cold water storage tank by the booster pump. The volume of water in the high-level storage tank is there to iron out the variations of flow into and out of the tank. At the beginning of the day (7 a.m.) the high-level storage tank will be full of water and ready to cope with the momentary simultaneous flow, assuming little is drawn off after 11 p.m. the previous evening. Do you agree with the logic here? You can see that the methodology is largely based on common sense.

Pump pressure developed must now be considered and will include: hydraulic resistance in the index circuit *plus* static lift *plus* discharge pressure at the ball valve (index terminal) in the high-level tank.

For a flow rate of 1.7 litres/s in 42 mm copper pipe to table X [*CIBSE Concise Handbook Section C*] the pressure drop is 416 Pa/m. If the index run is 50 m and taking an allowance for fittings of 15% on straight pipe, the hydraulic resistance is

$$416 \times 50 \times 1.15 = 24 \, \text{kPa}$$

Static lift is equivalent to:

$$h \times \rho \times g = 3 \times 16 \text{ storeys (to tank room)} \times 1000 \times 9.81$$
$$= 471 \, \text{kPa}$$

Recommended discharge pressure at the ball valve is 30 kPa. Pump pressure developed therefore is $24 + 471 + 30 = 525$ kPa. Net booster pump duty is 1.7 litres/s at 525 kPa. You will have noticed that the discharge pipe from the booster pump has been sized at 42 mm.

Note: If all the water is stored at high level and the booster pump is fitted directly to the rising main, the rate of flow required in 2 h to refill the tank, which now stores 10 800 litres of water, will be $10\,800/(2 \times 3600) = 1.5$ litres/s. This is added to the drinking water requirement of 1.2 litres/s, giving 2.7 litres/s.

Pump pressure developed will be 525 kPa *minus* the minimum mains pressure of 2 bar gauge *minus* the original hydraulic resistance at 1.7 litres/s *plus* the hydraulic resistance at 2.7 litres/s, which in 54 mm copper pipe to table X is 16 kPa. Thus pump pressure $= 525 - 200 - 24 + 16 = 317$ kPa.

Net booster pump duty is 2.7 litres/s at 317 kPa. Do you agree with this calculation? Clearly the pump now only has to generate *static* pressure equivalent to the static height *above* the minimum mains pressure.

Note: In either case the pressure developed by the booster pump is considerable, and not usually achieved by a single-stage centrifugal pump. Multistage pumps are commonly employed for this service.

(c)   The drinking water storage for the upper floors is calculated in two ways. In each case it is relatively small to insure against stagnation in the drinking water header and encourage a high rate of displacement. It is for this reason that the connection to the ball valve comes off the drinking water header.

One method of calculation is to allow 2 min storage: thus $2 \times 60 \times 1.2$ litres/s = 144 litres. The other method is to allow 4.5 litres/dwelling: thus $4.5 \times 4 \times 9$ floors = 162 litres. Size of drinking water header serving upper floors is 162 litres.

BOOSTED WATER: CONTROL BY WATER PRESSURE

This method of providing high-rise buildings with water supplies is more common, as it does not require electrical wiring from ground/basement where the booster pump is situated to the high-level tank room where the float switches are located in the storage tank and drinking water header.

There are a number of specialist pump manufacturers who offer water pressurization plant similar to that shown in Figure 9.2, where all the controls are located in the pressurization unit (see Figure 9.6).

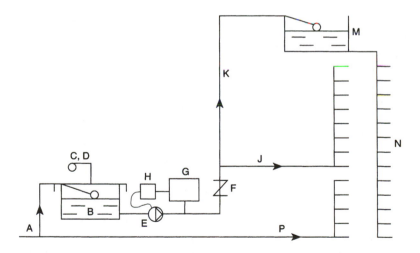

Figure 9.6   Boosted cold water, control by water pressure, valves omitted: A, mains water service; B, ground storage tank; C, D, filtered open vent and overflow; E, booster pump; F, recoil valve; G, pressure vessel; H, pressure switch; J, drinking water header to upper floors; K, boosted water to CWS tank; M, high-level cold water storage tank; N, cold water down-service serving urinals, WCs, basins, sinks etc.; P, mains drinking water to lower floors. E, G and H form the pressurization unit, which is supplied complete.

The cold water down service will require pressure reduction at intervals of five storeys to avoid excessive pressures at the draw-off points. The pressure vessel is partially gas-filled to reduce the number of pump operations. It is sized to hold the calculated quantity of drinking water. As water is drawn off from drinking water to the upper floors the drop in pressure activates the booster pump. The high-level cold water storage tank can be fitted with a delayed action ball valve so that the pump is activated at a predetermined low water level. This also assists in reducing the number of pump operations.

Case Study 9.5

The block of flats adopted in Case Study 9.4 is considered again for boosted water control by water pressure.
Design procedures include determination of:

(a) cold water storage;
(b) net pump duty;
(c) drinking water storage;
(d) size of pressure vessel.

SOLUTION

(a) Cold water storage remains the same, at 3600 litres at high level and 7200 litres at ground level.
(b) Pump flow rate remains the same at 1.7 litres/s. Pump pressure developed needs further consideration.
At no flow, cut-in pressure for the pump P_2 = static lift = 471 kPa gauge or 571 kPa absolute. At no flow, cut-out pressure P_3 for the pump = static lift plus one atmosphere differential:

$$471 + 100 = 571 \text{ kPa gauge or } 671 \text{ kPa absolute}$$

Initial pressure P_1 is taken as one atmosphere *below* cut-in pressure and is equal to 371 kPa gauge or 471 kPa absolute. Net pump duty is therefore design flow at pressure P_3 and will be 1.7 litres/s at 571 kPa gauge. This compares with 1.7 litres/s at 525 kPa gauge for control by water level.
(c) The reason for identifying the absolute pressures P_1, P_2 and P_3 in part (b) of the solution is that they are needed to size the pressure vessel. The process of pressurization is isothermal, and therefore Boyle's law can be adopted. From Chapter 4:

$$V_1 = \frac{P_2}{P_1} \left(\frac{P_3 E}{P_3 - P_2} \right)$$

where V_1 = volume of the vessel and E = drinking water storage requirement of 144 litres. The volume of the pressure vessel therefore

$$\frac{571}{471}\left(\frac{671 \times 144}{671 - 571}\right) = 1171 \text{ litres}$$

Note:

1. The difference in the cut-in and cut-out pressures is 1 atm to ensure sufficient differential to provide clear signals for pump operation.
2. In the determination of pressures P_2 and P_3, when the pump is operating against no flow, it is helpful to remember that there is no hydraulic loss and no pressure required for discharge at the ball valve.
3. The sum of the hydraulic loss and ball valve discharge pressure is less than 1 atmospheric differential, which is therefore taken to determine pressure P_3.

BOOSTED WATER: CONTINUOUS PUMP OPERATION

The continuously running pump system, which is a useful application for smaller systems, relies for its control upon careful pump selection and a small bleed from pump delivery to suction (Figure 9.7). Potable water and water for washing and sanitary use can be supplied in this way following approval from the water undertaking.

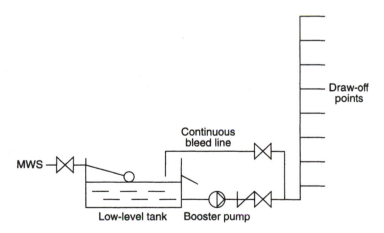

Figure 9.7 Continuously running booster pump. Tank requires sealed cover and filtered open vent and overflow.

DELAYED-ACTION BALL VALVES

To reduce the number of starts for booster pumps controlled by water/gas pressure, a delayed-action ball valve may be fitted to the mains water supply serving the high-level storage tank. There are various types on the market, but each ensures that the ball valve remains closed after the tank is full until the water level reaches a predetermined low point in the tank. This can have the effect of countering the need for sufficient storage in the event of an interruption of the water supply occurring at low water level in the tank. Advice should be sought from the water undertaking.

WATER SUPPLIES TO BUILDING IN EXCESS OF 20 STOREYS

Cold water storage

As a rule of thumb, cold water storage would be taken via a ground storage tank to intermediate storage tanks at 10 floor intervals (Figure 9.8).

Drinking water storage

The provision of potable water in high-rise buildings must insure against bacterial growth and contamination. This usually requires the use of sealed tanks, vessels or headers. For buildings over 20 storeys, ground storage may

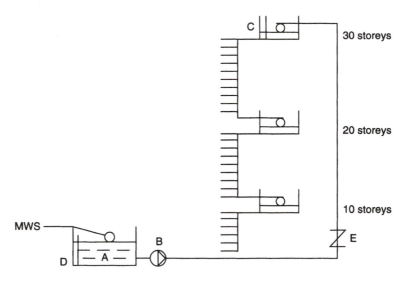

Figure 9.8 Boosted cold water above 20 storeys; A, ground storage tank; B, booster pump; C, float switch pump control; D, low-level float switch pump control; E, recoil valve.

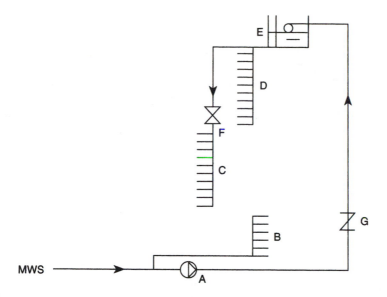

Figure 9.9 Boosted drinking water direct off the MWS, valves omitted: A, booster pump; B, drinking water to lower floors off rising main; C, drinking water to upper floors; D, drinking water to top floors; E, float switch control on pump; F, fixed pressure-reducing valve on drinking water supply to upper floors; G, recoil valve.

be advised, with further storage provided on the upper floors. Alternatively, the booster pump may be connected direct to the mains water supply to serve a storage vessel at roof level (Figure 9.9). The drinking water storage tank at roof level must be fitted with a sealed cover and filtered overflow and open vent.

Case Study 9.6

A 30-storey office block having a central core of toilet accommodation and basement allocation for water storage is at the feasibility design stage. Estimate the duty and power requirement for the boost pumps handling the domestic water services.

DATA

Occupancy is to be taken as 100 per floor except the ground floor, which consists of the entrance, reception and showcase booths. Floor heights are 3 m.

SOLUTION

Two water services are required here: for washing and sanitary use, and for drinking.

From the *CIBSE Guide [Concise Handbook Section G]*, offices without canteens require 40 litres/person domestic storage to cover a 24 h interruption of water supply. Total storage = $40 \times 100 \times 29 = 116\,000$ litres of water. If 8 h interruption is adequate for offices, the total storage = $116\,000 \times 8/24 = 38\,667$ litres.

This is equivalent to a mass of 39 tonnes of water! If two-thirds is stored at basement level and one-third is split between three tanks located at floor 10, floor 20 and the roof (Figure 9.8), the storage arrangements will be:

$$\text{Basement storage} = 38\,667 \times \frac{2}{3} = 25\,800 \text{ litres}$$

$$\text{10th floor, 20th floor and roof storage} = 38\,667 \times \frac{1}{3} \times \frac{1}{3} = 4300 \text{ litres each}$$

If the booster pump must refill these tanks in 4 h the rate of flow handled will be

$$4300 \times \frac{3}{4 \times 3600} = 0.9 \text{ litres/s}$$

Pressure developed by the booster pump will be static lift *plus* hydraulic loss *plus* discharge pressure at the index ball valve.

$$\text{Static lift} = 31 \times 3 \times 1000 \times 9.81 = 913 \text{ kPa}$$

Hydraulic loss for 0.9 litres/s in copper Table X tube [*CIBSE Concise Handbook Section C*], 42 mm bore is 133 Pa/m and allowing 15% for fittings on straight pipe:

$$\text{hydraulic loss} = 31 \times 3 \times 1.15 \times 133 = 14 \text{ kPa}$$

Recommended discharge pressure at the roof tank ball valve = 30 kPa.

$$\text{Pump pressure} = 913 + 14 + 30 = 957 \text{ kPa}$$

Estimated net pump duty is 0.9 litres/s at 957 kPa.

There are two equations for determining pump power:

$$\text{power} = Mgh = VFR \times \mathrm{d}p \quad (\mathbf{W})$$

where M = mass flow rate (kg/s), h = pump head (m), VFR = volume flow rate (m^3/s), and dp = pressure developed by the pump (Pa).

Adopting the second equation,

$$\text{power} = 0.0009 \times 962\,000 = 866\,\text{W}$$

Taking overall pump efficiency as 50%,

$$\frac{866}{0.5} = 1732\,\text{W}$$

The drinking water for the offices must now be addressed (see Figure 9.9). Assuming that the rising main can supply the first 10 floors, the minimum mains water pressure at ground level will need to be approximately 325 kPa. You should now confirm this minimum water pressure. The remaining floors must be served by a booster pump.

Normally one drinking water point is allocated to each toilet suite. One drinking water point will therefore be allocated per floor. If 10 DU are taken for each drinking water point to the upper floors, total DU = $20 \times 10 = 200$, and from Table 7.9 in Chapter 7 this is 0.8 litres/s. This is equivalent to $0.8/0.15 = 5$ drinking water points discharging simultaneously out of a total of 20, which we shall take as adequate. Do you agree with this decision?

If the booster pump is located in the mains water supply, the pressure developed will be: static lift *minus* lift provided by minimum mains water pressure *plus* hydraulic loss *plus* discharge pressure at the index ball valve:

$$\text{Static lift} = 913\,\text{kPa}$$

$$\text{static lift provided by mains water pressure} = 325\,\text{kPa}$$

Hydraulic loss for 0.8 litres/s in copper tube Table X and 35 mm bore [*CIBSE Concise Handbook Section C*] is 260 Pa/m, and allowing 15% for fittings on straight pipe:

$$\text{hydraulic loss} = 31 \times 3 \times 1.15 \times 260 = 28\,\text{kPa}$$

Recommended discharge pressure at the roof tank ball valve = 30 kPa.

$$\text{Pump pressure} = 913 - 325 + 28 + 30 = 646\,\text{kPa}$$

Estimated net pump duty is 0.8 litres/s at 646 kPa. Estimated pump powertaking an overall efficiency of 50% is 1034 W. Note the influence that the time to refill the tank has on the duty and power requirement of the booster pump.

Size of DW storage tank $= 0.8 \times 2 \times 60 = 96$ litres for 2 min storage

9.5   Variable speed pumps

Circulating pumps fitted with inverters can be used to advantage in boosted water systems. They operate using pressure transducers that convert mechanical variations into electrical signals via a logic controller. The duty pump attempts to keep the discharge pressure constant by varying the rotational speed as water flow varies resulting from changing consumption patterns during the day. A pressure vessel still forms part of the pressurization unit to damp out the frequency of changes in pressure that can occur in the system. This type of water booster set can cope with no flow conditions with the pressure transducer sensing increasing pressure and therefore stopping the pump.

There are two options that can be considered in system design for buildings over 10 storeys:

- Use of one booster set located at basement or ground floor level;
- The use of a booster set located at say ground, 7th and 14th floor (see Figure 9.10).

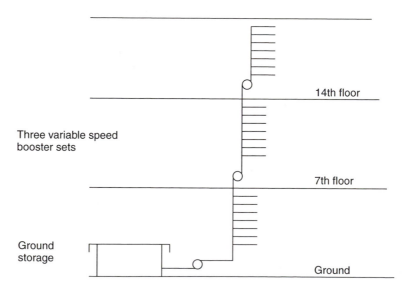

Figure 9.10   Showing the use of three variable speed booster sets for the provision of water services in high rise buildings.

There are advantages for the latter approach that requires three booster sets instead of one. The overall power consumption will be lower resulting from the smaller duties of the booster sets on the 7th and 14th floors and maximum discharge pressure at the draw off points will be limited to the static pressure of seven floors and this is equivalent to about 80 kPa. Full advantage of this limiting static pressure is gained if all the water is stored at ground level. If roof storage is required the cold water down service will be subject to the 20 storeys of static pressure but not the drinking water service that is delivered off the boosted main.

A single booster set serving a building of 20 storeys means that the static pressure at a draw off point on the ground floor will be in the region of 230 kPa. This is too high and pressure limiting valves will need to be fitted on the first 15 floors.

9.6   Chapter closure

You now have the skills to design a centralized HWS system requiring secondary pumping where the static head from the high-level storage tank is sufficient to provide static lift but insufficient to overcome the hydraulic losses in the secondary circuits. You are able to employ a pressurization unit for a system of centralized HWS where the cold water supply serving the HWS storage vessel has insufficient or no static head available.

You have been introduced to the special requirements for cold water supplies to high-rise buildings where the mains water service pressure is insufficient to provide the required static lift. The foregoing case studies provide you with the skills to identify the requirements and to design cold water services and drinking water services to multi-storey buildings.

It is important to remember to seek advice from the local water undertaking in respect of storage and pumping arrangements before proceeding with the design of water services within buildings.

10 Flues and draught

Nomenclature

CV	net calorific value (MJ/m^3, MJ/kg)
d	flue diameter (m)
f	frictional coefficient
g	gravitational acceleration (taken as 9.81 m/s^2 at sea level)
h	vertical height of the flue or stack (m)
k	velocity head loss factor
k_t	total velocity head loss factor
L	length of straight flue (m)
M_a	mass flow of air (kg/s)
M_{fg}	mass flow of flue gas (kg/s)
n	air ratio to fuel by volume
P_1	pressure at the base of the corresponding column of air (Pa)
P_2	pressure at the base of the column of flue gas (Pa)
T_1	absolute temperature of the flue gas at ambient conditions (K)
T_2	absolute temperature of the flue gas at mean flue gas temperature (K)
u	mean velocity (m/s)
$(VFR)_f$	volume flow rate of the fuel (m^3/s)
$(VFR)_{fg1}$	volume flow rate of the flue gas at ambient conditions (m^3/s)
$(VFR)_{fg2}$	volume flow rate of the flue gas at the mean flue gas temperature (m^3/s)
$(VFR)_{fd}$	volume flow rate of forced draught (m^3/s)
x	leakage %
Z	height above a datum (potential energy) (m) of flue gas
η	thermal efficiency
ρ_{fg1}	density of flue gas at ambient temperature (20 °C) (kg/m^3)
ρ_{fg2}	density of flue gas at the mean flue gas temperature (kg/m^3)

10.1 Introduction

Selecting the right flue for a particular boiler plant is a specialist's task and is usually done by the flue manufacturer in conjunction with the boiler manufacturer. The fossil fuel being burnt is also a factor since the combustion products of coal, fuel oil and natural gas from boiler plant above a rating of around 360 kW need to discharge, so that dispersal does not affect the area in the path of the prevailing wind until the products have been sufficiently diffused in the atmosphere. Flue height is therefore dependent on this and on the height of surrounding buildings, obstructions and the local terrain. Common combustion pollutants considered when calculating flue discharge heights include nitric oxide (NO), nitrogen dioxide (NO_2) and sulphur dioxide (SO_2).

The Acts which effectively cover chimney height include the Clean Air Act 1993, the Environmental Protection Act 1990 and the National Air Quality Strategy published by the Stationary Office CM3587 in March 1997. Recourse can be made to the Acts and the Local Authority in the process of determining flue height. Alternatively advice from a specialist flue manufacturer is worth pursuing. [*Flue manufacturers*, Selkirk Manufacturing Ltd.] Flues must be thermally well insulated so that they rapidly warm up and to ensure against condensation occurring in the products of combustion.

10.2   Generic formulae

- Boiler thermal efficiency:

$$\eta = \frac{\text{output}}{\text{input}} = \frac{\text{output}}{\text{CV} \times (\text{VFR})_f}$$

from which $(\text{VFR})_f$ can be evaluated.

- Properties of flue gas:
  Since burnt fuel content is minimal compared with the air supplied for combustion it is ignored and the flue gas is considered as air but only in respect of its density.

- Correcting for temperature:
  Flue gas density at flue gas temperature,

$$\rho_{fg2} = \rho_{fg1}\left(\frac{T_1}{T_2}\right)$$

- Flue gas:
  Flue gas volume flow rate at ambient temperature,

$$(\text{VFR})_{fg1} = (\text{VFR})_f(n + 1)$$

Flue gas volume flow rate at flue gas temperature,

$$(\text{VFR})_{fg2} = (\text{VFR})_{fg1}\left(\frac{T_2}{T_1}\right)$$

10.3 Natural draught

Natural draught (Figure 10.1) is dependent upon the mean temperature of the flue gas column in the stack and the temperature of a corresponding column of atmospheric air. The difference in density between the two columns along with the height of the flue allows the calculation of the natural draught available. The pressure at the base of each column forms the basis for the calculation of natural draught.

$$P_1 = h\rho_1 g \quad \text{Pa} \quad \text{and} \quad P_2 = h\rho_2 g \quad \text{Pa} \quad \text{where } P_1 > P_2$$

and, therefore, natural draught available

$$P_1 - P_2 = h(\rho_1 - \rho_2)g \quad \text{Pa}$$

The magnitude of the natural draught available increases with flue height h. Clearly at the point of plant start up there is little temperature difference and any draught created in the flue will depend upon the effect of wind speed at the flue terminal. It is therefore essential to ensure that the flue is well insulated so that it rapidly warms up allowing maximum natural draught to be reached quickly.

Practical flue diameter: To allow for a stationery flue gas film of some thickness around the inside perimeter of the stack practical diameter,

$$d = 1.4 \times \text{theoretical diameter}$$

Now

$$(\text{VFR})_{\text{fg2}} = \text{mean velocity} \times \text{cross sectional area}$$

$$= u \times \frac{\pi \,(\text{theoretical } d\,)^2}{4}$$

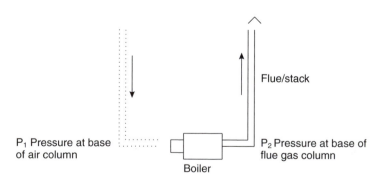

Figure 10.1 Natural draught.

from which the theoretical flue diameter d can be evaluated.

$$d = \sqrt{\frac{4 \, (\text{VFR})_{\text{fg2}}}{(u\pi)}}$$

Example 10.1

From the data relating to a natural draught flue connected to a gas fired boiler, determine:

(i)   the natural draught available when the plant is run up to the temperature;
(ii) the volume flow rate of flue gas;
(iii) the stack diameter.

Data

Stack height 9 m, ambient air density at 20 °C is 1.2 kg/m³, mean flue gas temperature 145 °C, boiler output 200 kW, calorific value of natural gas 39.4 MJ/m³, boiler thermal efficiency 80%, air fuel ratio by volume 10:1, mean flue gas velocity in the stack 7 m/s.

Solution

(i) $\rho_{\text{fg2}} = 1.2 \left(\dfrac{273 + 20}{273 + 145} \right) = 0.84 \, \text{kg/m}^3$

$$dP = 9(1.2 - 0.84)9.81 = 31.7 \, \text{Pa}$$

(ii) $\eta = \dfrac{\text{output}}{\text{input}}$

thus

$$0.8 = \frac{200}{39.4 \times 1000 \times (\text{VFR})_{\text{f}}}$$

from which

$$(\text{VFR})_{\text{f}} = \frac{200}{39.4 \times 1000 \times 0.8} = 0.006345 \, \text{m}^3/\text{s}$$

At 20 °C

$$(\text{VFR})_{\text{fg1}} = (\text{VFR})_{\text{f}}(n+1) = 0.006345(10+1) = 0.0698 \, \text{m}^3/\text{s}$$

At mean flue gas temperature,

$$(\text{VFR})_{\text{fg2}} = (\text{VFR})_{\text{fg1}} \left(\frac{T_2}{T_1} \right)$$

Thus at 145 °C

$$(\text{VFR})_{\text{fg2}} = \frac{0.0698(273 + 145)}{273 + 20} = 0.0996 \, \text{m}^3/\text{s}$$

(iii) $(\text{VFR})_{\text{fg2}} = u \times \dfrac{\pi \, (\text{theoretical } d)^2}{4}$

from which

$$d = \sqrt{\frac{4 \, (\text{VFR})_{\text{fg2}}}{u\pi}}$$

so

$$\text{theoretical } d = \sqrt{\frac{4 \times 0.0996}{7 \times \pi}} = 0.135 \, \text{m}$$

and

$$\text{practical stack diameter} = 1.4 \times 0.135 = 0.189 \, \text{m}$$

So practical stack diameter is 190 mm. It is likely that the commercially available flue for this application will be 200 mm diameter.

10.4   Draught stabilizers

Draught stabilizers (Figure 10.2) are fitted below the boiler smoke pipe connection to the flue to iron out fluctuations in natural draught caused by varying wind speed and direction at the flue terminal. They can also be used on boilers with forced draught. Their use does have the effect of reducing stack temperature due to the introduction of ambient air from the plant room. This is another reason for a well insulated flue to ensure against condensation.

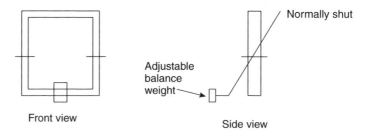

Figure 10.2   The draught stabilizer.

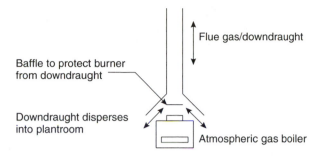

Figure 10.3 The draught diverter.

10.5   Draught diverters

Gas fired boilers with atmospheric burners can come with a draught diverter (Figure 10.3) that allows down draught from within the flue to disperse away from the burner jets. It also prevents excessive draught at the burner jets caused by high wind speed at the flue terminal by drawing air from the plant room.

10.6   Forced draught

Many fuel burners fitted to boiler plant come with an integral forced draught fan (Figure 10.4). This allows the combustion engineers to have control over the fuel/air atomization and control over the behaviour of the flame in the combustion chamber. The fan must be sized to handle the maximum air required at ambient temperature for combustion of the fuel and to overcome the frictional losses within the combustion chamber, flue gas passes within the boiler and this can include the frictional losses in the flue. There is positive pressure within the combustion chamber and a tendency for a small leakage outward into the plant room through joints.

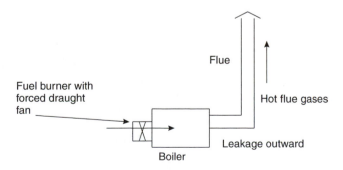

Figure 10.4 Forced draught.

Since the products of combustion are fan assisted no allowance for a still layer of flue gas on the inside walls of the flue is made in calculating the net flue diameter. At ambient temperature,

$$(VFR)_{fg1} = (VFR)_f (n + 1)(1 - x)$$

Example 10.2
A forced draught boiler operates on natural gas and on full load consumes 1200 l/min. Determine the stack diameter.

Data
Flue gas velocity 7 m/s, volumetric air fuel ratio 10:1, mean flue gas temperature 140 °C, air density at 20 °C is 1.2 kg/m³, leakage 3%.

Solution
At ambient conditions,

$$(VFR)_{fg1} = \left(\frac{1200}{60}\right)(10 + 1)(1 - 0.03) = 213.4 \, l/s$$

At 140 °C

$$(VFR)_{fg2} = 0.2134\left(\frac{273 + 140}{273 + 20}\right) = 0.3008 \, m^3/s$$

Stack diameter,

$$d = \sqrt{4 \times \frac{0.3008}{\pi \times 7}} = 0.234 \, m$$

So net stack diameter = 234 mm.

10.7 Induced draught

The induced draught (Figure 10.5) fan is located in the flue connection to the boiler and therefore handles the hot flue gases. It induces a negative pressure in the combustion chamber and therefore ensures against toxic gases entering the plant room as any leakage through joints is now inward. At ambient conditions,

$$(VFR)_{fg1} = (VFR)_f (n + 1)(1 + x)$$

Fan duty consists of the volume flow rate of the hot flue gases and the pressure developed to overcome the losses within the boiler and the stack.

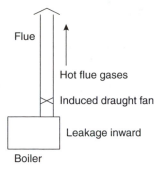

Figure 10.5   Induced draught.

Example 10.3
Using the same boiler and data in Example 10.2 find the stack diameter if
the plant is fitted with an induced draught fan instead of a forced draught
fan. Assume that the 3% leakage inward has no practical effect on flue
gas temperature.

Solution

$$(VFR)_{fg1} = \left(\frac{1200}{60}\right)(10 + 1)(1 + 0.03) = 226.6 \, litres/s$$

$$(VFR)_{fg2} = 0.2266\left(\frac{273 + 140}{273 + 20}\right) = 0.3194 \, m^3/s$$

$$\bar{d} = \sqrt{4 \times \frac{0.3194}{\pi} \times 7} = 0.241 \, m$$

So net stack diameter = 241 mm.

10.8   Determination of fan duty

FORCED DRAUGHT FANS AND INDUCED DRAUGHT FANS

At ambient temperature the air handled by the fan,

$$(VFR)_{fd} = n\,(VFR)_f$$

Alternatively ambient air handled by the fan,

$$(VFR)_{fd} = \left(\frac{n}{n+1}\right)(VFR)_{fg1}$$

where at ambient conditions,

$$(\text{VFR})_{fg1} = (\text{VFR})_{fg2}\left(\frac{T_1}{T_2}\right)$$

Pressure developed by the fan = flue gas loss in boiler
+ flue gas loss in the flue

Table 10.1 gives velocity pressure loss values for typical flue fittings. The losses due to friction in the flue can be determined from the Bernoulli equation. (*Heat and Mass Transfer in Building Services Design*, Spon Press)

$$Z_1 + \left(\frac{P_1}{\rho g}\right) + \frac{u_1^2}{2g} = Z_2 + \left(\frac{P_2}{\rho g}\right) + \frac{u_2^2}{2g} + \text{Losses}$$

where P_1 is flue gas pressure at the base of the stack Pa gauge, P_2 is atmospheric pressure at the terminal = zero gauge, u_1 is flue gas velocity in the stack, u_2 is flue gas velocity as it leaves the terminal, and

$$u_1 = u_2.$$

So

$$\frac{P_1}{\rho g} = (Z_1 - Z_2) + \text{losses} \quad \text{(metres of flue gas)}$$

Since the flue gas is made buoyant by its high temperature, static lift $(Z_1 - Z_2)$ does not have to be provided by the fan.
Thus

$$\frac{P_1}{\rho g} = \text{loss in the straight flue} + \text{loss in flue fittings}$$

Therefore

$$\frac{P_1}{\rho g} = \left(\frac{4fLu^2}{2gd}\right) + \left(\frac{k_t u^2}{2g}\right) \quad \text{(metres of flue gas)}$$

and

$$P = \left(\left(\frac{4fLu^2}{2gd}\right) + \left(\frac{k_t u^2}{2g}\right)\right)\rho g \quad \text{Pa}$$

Example 10.4
A boiler is fitted with a pressure jet oil burner and the pressure loss in the flue gas passes within the boiler is 200 Pa at full load. The air fuel ratio by mass is 16.8 : 1. Find the duty of the forced draught fan fitted to the fuel burner.

Data
From Table 10.1 velocity head loss factors for flue fittings: terminal 1.4, 2 obtuse bends 0.2 each, 1 mitred bend 1.3. Straight length of flue 10 m, mean flue gas velocity 8 m/s, frictional coefficient 0.006. Mean flue gas temperature 140 °C, flue diameter 300 mm, air density at 20 °C is 1.2 kg/m^3, fuel oil density at 20 °C is 930 kg/m^3.

Solution
At 140 °C

$$(\text{VFR})_{\text{fg2}} = \frac{u\pi d^2}{4} = 8 \times \pi \times \frac{(0.3)^2}{4} = 0.5655 \, \text{m}^3/\text{s}$$

$$\text{Volume of air per kg of oil} = \frac{16.8}{1.2} = 14 \, \text{m}^3$$

$$\text{Volume of oil per kg} = \frac{1}{930} = 0.001075 \, \text{m}^3$$

VFR of air in flue gas at 140 °C

$$= 0.5655 \left(\frac{n}{n+1} \right)$$

$$= 0.5655 \left(\frac{14}{14 + 0.001075} \right) = 0.5655 \, \text{m}^3/\text{s}$$

Note the effect a volumetric ratio has on the term $(n+1)$.
   VFR of air at 20 °C

$$= 0.5655 \left(\frac{273 + 20}{273 + 140} \right) = 0.4012 \, \text{m}^3/\text{s}$$

Table 10.1   Velocity head loss factors

Fitting	k factor
135° obtuse bend	0.2
90° square mitred bend	1.3
135° obtuse tee	0.5 from branch
	0.3 from main section
Terminal (dependent upon type)	1.4–2.4

Thus

$$(VFR)_{fd} = 0.4012 \, \text{m}^3/\text{s}$$

Flue gas density,

$$\rho_2 = 1.2 \left(\frac{273 + 20}{273 + 140} \right) = 0.85 \, \text{kg/m}^3$$

Total head loss factor,

$$k_t = 1.4 + 1.3 + 0.4 = 3.1$$

Pressure loss in the stack,

$$P = \left(\left(\frac{4 \times 0.006 \times 10 \times 8^2}{2g \times 0.3} \right) + \left(3.1 \times \frac{8^2}{2g} \right) \right) \times 0.85 \times g$$
$$= (2.61 + 10.112) \times 0.85 \times g$$
$$= 106 \, \text{Pa}$$

Total pressure developed by the force draught fan = boiler + flue
$$= 200 + 106 = 306 \, \text{Pa}$$

Net fan duty therefore will be $0.4012 \, \text{m}^3/\text{s}$ at $306 \, \text{Pa}$. Note the forced draught fan handles atmospheric air at ambient temperature.

Example 10.5
Using the data in example 10.4, determine the duty of an induced draught fan assuming that the oil burner is of the nonaspirated type and mean flue gas temperature is $140 \,°\text{C}$.

Solution
At $140 \,°\text{C}$ $(VFR)_{fg2} = 0.5655 \, \text{m}^3/\text{s}$, pressure loss in stack $dP = 106 \, \text{Pa}$

Total pressure developed by the induced draught fan = boiler + flue
$$= 200 + 106 = 306 \, \text{Pa}$$

Net fan duty therefore will be $0.5655 \, \text{m}^3/\text{s}$ at $306 \, \text{Pa}$. Note the induced draught fan handles the products of combustion at $140 \,°\text{C}$.

Example 10.6

A boiler having a rated output of 150 kW is fired by natural gas from a pressure jet burner. The boiler efficiency is 84% and the calorific value of natural gas is 39.4 MJ/m^3. From the data determine the stack diameter and the net duty of the forced draught fan. Ignore leakage.

Data

Pressure loss in the flue gas passes in the boiler 220 Pa, volumetric air fuel ratio 11:1, height of flue 6 m, mean flue gas temperature 145 °C, mean flue gas velocity 7 m/s, coefficient of friction in the stack 0.005, velocity head loss factors for stack fittings include 1 mitred bend 1.3 and 1 terminal 1.9, density of air at 20 °C is 1.2 kg/m^3.

Solution

Gas flow rate at ambient conditions,

$$(\text{VFR})_f = \frac{150}{39\,400 \times 0.84} = 0.0045323 \, \text{m}^3/\text{s}$$

At 145 °C flue gas,

$$(\text{VFR})_{fg2} = 0.0045323(11+1)\left(\frac{273+145}{273+20}\right)$$
$$= 0.0776 \, \text{m}^3/\text{s}$$

from which

$$\text{stack diameter,} \ d = \sqrt{\frac{4 \times 0.0776}{\pi \times 7}} = 0.119 \, \text{m}$$

Net stack diameter, d is 119 mm. Ambient air handled by the forced draught fan,

$$(\text{VFR})_{fd} = 0.0045323 \times 11 = 0.049855 \, \text{m}^3/\text{s}$$

$$\text{Flue gas density,} \ \rho_2 = 1.2\left(\frac{273+20}{273+418}\right) = 0.84 \, \text{kg/m}^3$$

$$\text{Pressure loss in flue} = \left(\left(\frac{4fLu^2}{2gd}\right) + \left(\frac{k_t u^2}{2g}\right)\right)\rho g$$
$$= \left(\left(\frac{4 \times 0.005 \times 6 \times 7^2}{2g \times 0.119}\right) + \left(\frac{3.2 \times 7^2}{2g}\right)\right) \times 0.84 \times g$$
$$= (2.5184 + 7.9918) \times 0.84 \times 9.81$$
$$= 86.6 \, \text{Pa}$$

Total pressure loss in system is $86.6 + 220 = 307$ Pa and the net duty of forced draught fan is $0.05 \, \text{m}^3/\text{s}$ at 307 Pa.

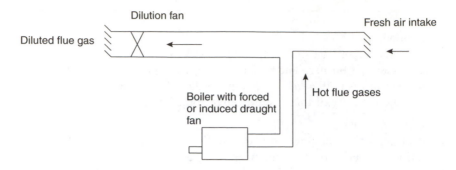

Figure 10.6   Diluted draught.

10.9   Diluted draught

The use of diluted draught (Figure 10.6) when permitted avoids the need for a flue rising to at least the height of the building that the heating plant is serving. Fresh air is drawn in by the diluting fan and mixed with the flue gas for discharge at a minimum of 3 m from ground. It is essential that the exhaust louvre is set to discharge at an angle of 30° above the horizontal to avoid the diluted flue gas from being entrained down the external wall thus preventing the dispersal of pollutants and causing condensation on the wall surface. A drain pipe is required from the underside of the diluted flue before it exits from the building to allow condensed vapour which will be slightly acidic to discharge. The fan is sized to carry the diluted products and overcome the pressure loss from the fresh air inlet to the dilution outlet. The forced draught fan on the fuel burner accounts for the losses through the boiler and boiler flue.

Example 10.7
A boiler fitted with a medium pressure burner operates on natural gas with an output of 300 kW and the products of combustion are removed using a diluted flue as shown in Figure 10.6. The temperature of the diluted products shall not exceed 40 °C at the point of discharge to atmosphere. From the data size the three flue sections.

Data
Net calorific value 39 MJ/m³, boiler efficiency 86%, volumetric air fuel ratio 10:1, flue gas temperature 140 °C, flue gas velocity 5 m/s, air density 1.2 kg/m³ at an ambient temperature of 20 °C, temperature of entering fresh diluting air 15 °C. Ignore leakage of combustion products from the boiler.

Solution

(a)   Boiler flue diameter

$$\eta = \frac{\text{output}}{CV \times (VFR)_f}$$

so

$$(VFR)_f = \frac{300}{39 \times 1000 \times 0.86} = 0.008945 \, \text{m}^3/\text{s}$$

At 20 °C

$$(VFR)_{fg1} = 0.008945 \times (10 + 1) = 0.09839 \, \text{m}^3/\text{s}$$

At 140 °C

$$(VFR)_{fg2} = 0.09839 \times \frac{413}{293} = 0.13869 \, \text{m}^3/\text{s}$$

$$\text{From } VFR = u \times \frac{\pi d^2}{4}, \qquad d = \sqrt{\frac{4(VFR)}{u \times \pi}}$$

$$\text{So } d = \sqrt{4 \times \frac{0.13869}{5 \times \pi}} = 0.188 \, \text{m}$$

Thus the boiler flue pipe net d is 190 mm.

(b)   Fresh air duct diameter

In order to find the quantity of fresh air required to keep the diluted products at no more than 40 °C a mass heat balance can be used at the junction of the fresh air duct and the boiler flue, thus:

Heat gain to the fresh air = heat loss from the boiler flue products

$$M_a \times C \times (40 - 15) = M_{fg} \times C \times (140 - 40)$$

The specific heat capacity C of the air/flue gas can be taken as equal, as the fuel product fraction in the flue gas is small and therefore it can be ignored.

At 20 °C

$$(VFR)_{fg1} = 0.09839 \, \text{m}^3/\text{s}$$

so

$$M_{fg1} = 0.09839 \times 1.2$$
$$= 0.118 \, \text{kg/s}$$

If M_a = the mass of fresh air, then substituting values into the mass heat balance,

$$M_a(40 - 15) = 0.118(140 - 40)$$

So

$$25 \times M_a = 100 \times 0.118 = 11.8$$

from which the mass flow of fresh air, (M_a) is 0.472 kg/s. Air density at 15 °C,

$$\rho = 1.2\left(\frac{293}{288}\right) = 1.221 \, \text{kg/m}^3$$

At 15 °C,

$$(\text{VFR})_a = \frac{M_a}{\rho} = \frac{0.472}{1.221} = 0.387 \, \text{m}^3/\text{s}$$

The net diameter of the fresh air duct,

$$d = \sqrt{\frac{4 \times 0.387}{5 \times \pi}} = 0.314 \, \text{m or } 314 \, \text{mm}$$

Note the mass ratio of fresh air to flue gas is 0.472/0.118 = 4:1.
   (c)   Diluted flue gas duct diameter
The density of the diluted products at 40 °C,

$$\rho = 1.2\left(\frac{293}{313}\right) = 1.123 \, \text{kg/m}^3$$

At 40 °C the volume flow of the diluted products,

$$(\text{VFR})_d = \frac{M}{\rho} = \frac{0.118 + 0.472}{1.123} = 0.525 \, \text{m}^3/\text{s}$$

The net diameter of the duct carrying the diluted products,

$$d = \sqrt{\frac{4 \times 0.525}{5 \times \pi}}$$

So $d = 0.366$ m or 366 mm.

Summary
Boiler flue pipe diameter is 190 mm, fresh air duct diameter is 314 mm and diluted flue gas duct section is 366 mm.
   Note the mean velocity for each duct section was taken as 5 m/s.

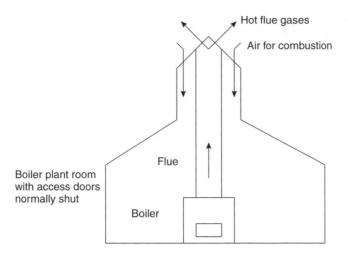

Figure 10.7 Balanced draught.

10.10 Balanced draught

The principle relies on the air for combustion being located adjacent to the point where the products of combustion are discharged (Figure 10.7). This ensures that the pressure effects and wind speed and direction at this point have the same effect on both fluid flows allowing equal pressures in the boiler combustion chamber. The vertical balanced flue offered by one manufacturer makes an architectural feature of the roof terminal (*Monodraught Ltd.*). An advantage with this type of draught provision is that the plant room can be located within the building away from an external wall.

10.11 Chapter closure

You have now completed an introduction to alternative ways of discharging the products of combustion from boiler plant, flue sizing and determination of net duties for the forced draught and induced draught fan. Flue design including the construction and determination of minimum height is a specialist function requiring the need to keep within the appropriate Acts and Regulations. Recourse is therefore needed to further reading of the Acts and of specialist manufacturers' literature.

The reader should note that if stack losses are not included in the fan pressure, stack diameter should be increased by 20%.

11 Combustion of fossil fuels

Nomenclature

CEI	carbon emission index (kg/m^2)
CV	net calorific value (MJ/m^3, MJ/kg)
DD	corrected annual Degree Days
dt	design indoor to outdoor temperature difference (K)
EH	equivalent hours at full load (h)
M_c	mass flow rate of coal (kg/s)
M_g	mass flow rate of gas (kg/s)
M_o	mass flow rate of oil (kg/s, kg/h)
T	absolute temperature (K)
$(VFR)_g$	volume flow rate of gas (m^3/s)
ρ	density (kg/m^3)
η	efficiency

11.1 Introduction

The building services engineer is party to the burning of fossil fuels and therefore takes some responsibility in ensuring that they are burnt efficiently by selecting the right plant and space heating system for the building type and use to which it is put. Achieving minimum emissions of carbon dioxide gas from the boiler plant into the atmosphere is part of the brief of the building services engineer. A working knowledge of how to determine carbon dioxide emissions from the boiler plant, burning fossil fuels such as methane gas, natural gas, oil and coal is therefore part of the engineer's portfolio. The services engineer may also be required to estimate the annual fuel consumption for a proposed building or to investigate the annual fuel consumption of an existing plant. [*Energy management and operating cost in buildings*, Spon Press.] This may include determination of the annual emission of carbon dioxide gas and by how much this emission can be reduced by good housekeeping or the installation of new state-of-the-art plant. There is therefore a new impetus implied now in the Building Regulations to understand the principles of the combustion of fossil fuels.

11.2 Chemical equations

This section introduces three tables relating to the constituents of some fossil fuels and demonstrates how a chemical equation for the complete combustion of a constituent of a fuel is developed.

- Table 11.1 lists those elements and compounds associated with the common fossil fuels by notation, atomic mass, and where appropriate, molecular mass.
- Table 11.2 lists the molecular mass of some common hydrocarbons in fossil fuels and shows how molecular mass is calculated from the product of atomic mass and the number of atoms in the element.
- Table 11.3 lists the combustion equations for the typical constituents of methane, and natural gas.

11.3 The chemical reference point

Hydrogen is adopted as the datum or reference point in chemistry. The element hydrogen has the notation H_2. The full notation for the element hydrogen is $1H_2$ where the number 1 denotes that hydrogen has one molecule

Table 11.1 Some elements, compounds and hydrocarbons

	Symbol	Atomic mass	Molecular mass
Elements			
Hydrogen	H_2	1	2
Carbon	C	12	12
Oxygen	O_2	16	32
Sulphur	S	32	32
Nitrogen	N_2	14	28
Argon	A	40	40
Compounds			
Water vapour	H_2O	–	18
Carbon monoxide	CO	–	28
Carbon dioxide	CO_2	–	44
Sulphur dioxide	SO_2	–	64
Hydrocarbons			
Methane	CH_4	–	16
Ethylene	C_2H_4	–	28
Hexane	C_6H_{14}	–	86
Heptane	C_7H_{16}	–	100
Ethane	C_2H_6	–	30
Butane	C_4H_{10}	–	58
Pentane	C_5H_{12}	–	72
Propane	C_3H_8	–	44

Table 11.2   Calculation of molecular mass of some hydrocarbons

Hydrocarbon	Symbol	Calculation	Molecular mass
Methane	CH_4	$(12 \times 1) + (1 \times 4)$	16
Ethylene	C_2H_4	$(12 \times 2) + (1 \times 4)$	28
Hexane	C_6H_{14}	$(12 \times 6) + (1 \times 14)$	86
Heptane	C_7H_{16}	$(12 \times 7) + (1 \times 16)$	100
Ethane	C_2H_6	$(12 \times 2) + (1 \times 6)$	30
Butane	C_4H_{10}	$(12 \times 4) + (1 \times 10)$	58
Pentane	C_5H_{12}	$(12 \times 5) + (1 \times 12)$	72
Propane	C_3H_8	$(12 \times 3) + (1 \times 8)$	44

and the subscript 2 identifies the fact that hydrogen has two atoms. (Similarly the element carbon has the notation C whereas its full notation is $1C_1$.)

The *atom* is the smallest part of an element that can take part in a chemical reaction. The *molecule* is the smallest part of a substance that can exist in a free state while retaining its original properties. The *molecular mass* of a gaseous substance is the product of the mass of two standard volumes of hydrogen and the density of the substance.

Table 11.3   Combustion equations for some hydrocarbons

Hydrocarbon	Symbol	Combustion equation
Methane	CH_4	$CH_4 + 2O_2 \rightarrow CO_2 + 2H_2O$ $16 + 64 \rightarrow 44 + 36$
Ethylene	C_2H_4	$C_2H_4 + 3O_2 \rightarrow 2CO_2 + 2H_2O$ $28 + 96 \rightarrow 88 + 36$
Hexane	C_6H_{14}	$2C_6H_{14} + 19O_2 \rightarrow 12CO_2 + 14H_2O$ $\{(2 \times 12 \times 6) + (2 \times 1 \times 14)\} +$ $(19 \times 32) \rightarrow (12 \times 44) + (14 \times 18)$ $172 + 608 \rightarrow 528 + 252$
Heptane	C_7H_{16}	$C_7H_{16} + 11O_2 \rightarrow 7CO_2 + 8H_2O$ $100 + 352 \rightarrow 308 + 144$
Ethane	C_2H_6	$2C_2H_6 + 7O_2 \rightarrow 4CO_2 + 6H_2O$ $\{(2 \times 12 \times 2) + (2 \times 1 \times 6)\} +$ $(7 \times 32) \rightarrow (4 \times 44) + (6 \times 18)$ $60 + 224 \rightarrow 176 + 108$
Butane	C_4H_{10}	$2C_4H_{10} + 13O_2 \rightarrow 8CO_2 + 10H_2O$ $\{(2 \times 12 \times 4) + (2 \times 1 \times 10)\} +$ $(13 \times 32) \rightarrow (8 \times 44) + (10 \times 18)$ $116 + 416 \rightarrow 352 + 180$
Pentane	C_5H_{12}	$C_5H_{12} + 8O_2 \rightarrow 5CO_2 + 6H_2O$ $72 + 256 \rightarrow 220 + 108$
Propane	C_3H_8	$C_3H_8 + 5O_2 \rightarrow 3CO_2 + 4H_2O$ $44 + 160 \rightarrow 132 + 72$

The volume of 1 kg of hydrogen at $0\,°C$ and 1 standard atmosphere $(101\,325\,Pa)$ is $11.14\,m^3$.

Thus at $0\,°C$

$$\text{molecular mass} = 2 \times 11.14 \times \rho = 22.28\rho$$

and therefore at $0\,°C$ the density of a substance $\rho = \text{molecular mass}/22.28$. For example, the density of oxygen at $0\,°C$ is $32/22.28 = 1.436\,kg/m^3$, and the density of oxygen at the ambient temperature $(15\,°C)$,

$$\rho_2 = \rho_1 \left(\frac{T_1}{T_2}\right)$$

$$= 1.436\left(\frac{273}{288}\right)$$

$$= 1.361\,kg/m^3$$

11.4   Approximate analysis of dry air

Table 11.4 lists the constituents of dry air by mass and identifies its molecular mass. The constituents of carbon dioxide, hydrogen and argon are rounded with that for oxygen to 23%. Thus by mass the constituents of dry air are taken as 23% oxygen and 77% nitrogen. By volume the constituents are 21% oxygen and 79% nitrogen. Thus the ratio of dry air to oxygen by mass is 100/23 and the ratio of dry air to oxygen by volume is 100/21. The average molecular mass of dry air is shown in Table 11.5 to be 29. As dry air is a compound it does not have an atomic mass. From Section 11.3 the density of dry air at $0\,°C$

$$\rho_1 = \frac{29}{22.28} = 1.3016\,kg/m^3$$

and, therefore, the density of dry air at ambient conditions,

$$\rho_2 = 1.3016\left(\frac{273}{288}\right)$$

$$= 1.234\,kg/m^3$$

Table 11.4   Approximate analysis of dry air. Average molecular mass $= 28.97$

Dry air constituent	% by mass	Molecular mass	Product
Nitrogen	78.03	28	2185.84
Oxygen	20.99	32	672.68
Carbon dioxide	0.03	44	1.32
Hydrogen	0.01	2	0.02
Argon	0.94	40	37.60
		28.97	2897.46

Table 11.5 Solution to example 11.1

Constituent	% by volume	% by volume × molecular mass	% by mass
CH_4	92.6	$92.6 \times 16 = 1482$	$1482/1720 = 86$
C_2H_6	3.6	$3.6 \times 30 = 108$	$108/1720 = 6.3$
C_3H_8	0.8	$0.8 \times 44 = 35.2$	$35.2/1720 = 2$
C_4H_{10}	0.3	$0.3 \times 58 = 17.4$	$17.4/1720 = 1$
N_2	2.6	$2.6 \times 28 = 73$	$73/1720 = 4.2$
CO_2	0.1	$0.1 \times 44 = 4.4$	$4.4/1720 = 0.25$
		Totals 1720	99.75

11.5   Development of some common combustion equations

The following shows how the equations for the complete combustion by mass of the elements hydrogen, carbon and sulphur from Table 11.1 are formed.

$$2H_2 + O_2 \rightarrow 2(H_2O) \quad (2 \times 1 \times 2) + (16 \times 2) \rightarrow (2 \times 18) \quad 4 + 32 \rightarrow 36$$
$$C + O_2 \rightarrow CO_2 \quad (12 \times 1) + (16 \times 2) \rightarrow 44 \quad 12 + 32 \rightarrow 44$$
$$S + O_2 \rightarrow SO_2 \quad (32 \times 1) + (16 \times 2) \rightarrow 64 \quad 32 + 32 \rightarrow 64$$

The determination of the molecular mass of some hydrocarbons is shown in Table 11.2. Note that the hydrocarbons are made up from the addition of the carbon and hydrogen elements.

11.6   Constituents of a fossil fuel by mass and by volume

The chemical equations given in Tables 11.2 and 11.3 are by mass. Thus for the complete combustion of carbon to carbon dioxide $C + O_2 \rightarrow CO_2$ implies that 12 kg of carbon requires 32 kg of oxygen and produces 44 kg of carbon dioxide. This may also be expressed as

$$1\,\text{kg}\,C + \left(\frac{32}{12}\right)\text{kg}\,O_2 \rightarrow \left(\frac{44}{12}\right)\text{kg}\,CO_2$$

Furthermore,

$$1\,\text{kg}\,C + 2.67\,\text{kg}\,O_2 \rightarrow 3.67\,\text{kg}\,CO_2$$

The percentage analysis of the constituents of fossil fuels can be expressed by mass or by volume. It is helpful to be able to convert a gaseous fuel for example whose percentage analysis of its constituents may be given by volume to those constituents given by mass.

Example 11.1
The constituents of natural gas by volume consist of 92.6% methane, 3.6% ethane, 0.8% propane, 0.3% butane, 2.6% nitrogen and 0.1% carbon dioxide. Find the percentage analysis of the constituents by mass.

Solution
Table 11.5 provides the solution giving in the final column the constituents of natural gas by mass.

11.7   Complete combustion of a fossil fuel

The theoretical quantity of air required for complete combustion of a fuel is referred to as the stoichiometric air quantity. Complete combustion relies on intimate mixing of the oxygen in the air with the fuel. Since oxygen represents only 23% by mass and the mixing process in the combustion chamber is never total, excess air is normally required to ensure complete combustion.

A rich air fuel mix identifies a low air fuel ratio. A weak air fuel mix denotes a high air fuel ratio and hence a high amount of excess air. A rich air fuel mix induces carbon monoxide CO in the combustion products. This is poisonous and is the result of the presence of unburnt carbon. A weak air fuel mix increases the amount of carbon dioxide in the combustion products. The efficiency of combustion is therefore a function of the degree to which CO_2, CO and O_2 are present in the products of combustion.

Example 11.2
Methane gas has the following constituents by mass: 96% methane, 2.5% nitrogen and 1.5% carbon dioxide.

(a)   Find the stoichiometric volume of air for complete combustion of $1\,m^3$ of methane gas.
(b)   Find the actual volume of air assuming 20% excess is needed for complete combustion.
(c)   Find the volumetric air fuel ratio based upon the actual volume of air required.

Solution
Nitrogen and carbon dioxide pass through the combustion process unaffected. From Table 11.3, 0.96 kg of methane requires $(0.96 \times 64/16)$ kg of oxygen = 3.84 kg/kg of fuel. From Section 11.3, the density of oxygen at ambient temperature is $1.361\,kg/m^3$. Volume of oxygen required at ambient conditions is $3.84/1.361 = 2.82\,m^3/kg$ of fuel. From Section 11.4,

the ratio of dry air to oxygen by volume is 100/21. Thus the stoichiometric volume of air is $2.82 \times 100/21 = 13.4355 \, m^3/kg$ of fuel. From Section 11.3,

$$\text{density} = \frac{\text{molecular mass}}{22.28}$$

Determining the density of methane gas at $0\,°C$ by proportion using Table 11.1

$$\rho = \left(\frac{16}{22.28}\right)0.96 + \left(\frac{28}{22.28}\right)0.025 + \left(\frac{44}{22.28}\right)0.015 = 0.75045 \, kg/m^3$$

Density of the fuel at ambient conditions is $0.75045(273/288) = 0.711 \, kg/m^3$.

(a)   Stoichiometric volume of air is $13.4355 \times 0.711 = 9.56 \, m^3/m^3$ of fuel.
(b)   Volume of air actually supplied is $9.56 \times 1.2 = 11.47 \, m^3/m^3$ of fuel.
(c)   The volumetric air fuel ratio will be 11.47:1.

Example 11.3
Natural gas has the constituents by mass as given in the solution to Example 11.1.

(d)   Find the stoichiometric volume of air for complete combustion of one cubic metre of natural gas.
(e)   Find the actual volume of air assuming 20% excess is needed for complete combustion.
(f)   Find the volumetric air fuel ratio based upon the actual volume of air required.

Solution
Note: The constituents of nitrogen and carbon dioxide play no part in the combustion process and therefore pass through unaffected.
     From Table 11.3,

$$0.86 \, kg \text{ methane requires} \quad (64/16) \times 0.86 \text{ of oxygen} = 3.44 \, kg$$

$$0.063 \, kg \text{ ethane requires} \quad (224/60) \times 0.063 \text{ of oxygen} = 0.2352 \, kg$$

$$0.02 \, kg \text{ propane requires} \quad (160/44) \times 0.02 \text{ of oxygen} = 0.0727 \, kg$$

$$0.01 \, kg \text{ butane requires} \quad (416/116) \times 0.01 \text{ of oxygen} = 0.03586 \, kg$$

$$\text{Total} = 3.7838 \, kg/kg \text{ fuel}$$

From Section 11.3, the density of oxygen at ambient conditions is 1.361 kg/m^3. Volume of oxygen is $3.7838/1.361 = 2.78$ m^3/kg fuel. The stoichiometric volume of air is $(2.78)100/21 = 13.24$ m^3/kg fuel. From Section 11.3,

$$\text{density} = \frac{\text{molecular mass}}{22.28}$$

Determining the density of natural gas at 0 °C by proportion using Table 11.1,

$$\left(\frac{16}{22.28}\right) \times 0.86 + \left(\frac{30}{22.28}\right) \times 0.063 + \left(\frac{44}{22.28}\right) \times 0.02$$
$$+ \left(\frac{58}{22.28}\right) \times 0.01 + \left(\frac{28}{22.28}\right) \times 0.042 + \left(\frac{44}{22.28}\right) \times 0.0025$$
$$= 0.8256 \,\text{kg/m}^3 \text{ at } 0\,°\text{C}$$

Density of natural gas at ambient conditions is $0.8256 \times 273/288 = 0.7826$ kg/m^3.

(d) The stoichiometric volume of air is $13.24 \times 0.7826 = 10.36$ m^3/m^3 fuel.

(e) Actual air supplied is $10.36 \times 1.2 = 12.43$ m^3/m^3 fuel.

(f) Volumetric air fuel ratio is 12.43:1.

Example 11.4

Light grade fuel oil has the following constituents by mass: carbon 86.2%, hydrogen 12.4%, sulphur 1.4%.

(g) Find the stoichiometric mass of air for complete combustion of one kilogram of the oil.

(h) Find the actual mass of air assuming 30% excess is needed for complete combustion.

(i) Find the air fuel ratio by mass based upon the actual mass of air required.

Solution

(g) From Section 11.5,

0.862 kg carbon requires $(32/12) \times 0.862$ of oxygen $= 2.2987$ kg
0.124 kg hydrogen requires $(32/4) \times 0.124$ of oxygen $= 0.992$ kg
0.014 kg sulphur requires $(32/32) \times 0.014$ of oxygen $= 0.014$ kg
$$\text{Total} = 3.3047\,\text{kg/kg fuel}$$

From Section 11.4, the ratio of dry air to oxygen by mass is 100/23. The stoichiometric mass of air for complete combustion of the fuel is $3.3047 \times 100/23 = 14.37\,\text{kg/kg}$ fuel.

(h)   The actual air supplied is $14.37 \times 1.3 = 18.68\,\text{kg/kg}$ fuel. Note percentage excess air is increased with fuel oil which when atomized needs more air to burn completely than in the case of gas.

(i)   The air fuel ratio by mass is 18.68:1.

Note: The actual air supplied can be expressed in m^3/kg fuel by dividing by air density at ambient conditions of 15 °C, thus $18.68/1.234 = 15.14\,\text{m}^3/\text{kg}$ fuel.

Example 11.5
Medium Rank coal has the following constituents by mass: 81.8% carbon, 4.9% hydrogen, 4.4% oxygen, 1.8% nitrogen, 1.9% sulphur and 5.2% mineral matter.

(j)   Find the stoichiometric mass of air for complete combustion of 1 kg of the coal.

(k)   Find the actual mass of air assuming 50% excess is needed for complete combustion.

(l)   Find the air fuel ratio by mass based upon the actual mass of air required.

Solution
Note: Nitrogen plays no part in the combustion process. Mineral matter remains behind as ash.

(j)   From Section 11.5,

0.818 kg carbon requires $(32/12) \times 0.818$ of oxygen $= 2.1813\,\text{kg}$
0.049 kg hydrogen requires $(32/4) \times 0.049$ of oxygen $= 0.392\,\text{kg}$
0.019 kg sulphur requires $(32/32) \times 0.019$ of oxygen $= 0.019\,\text{kg}$

$$\text{Total} = 2.4923\,\text{kg/kg fuel}$$
$$\text{less the oxygen present in the fuel} = 0.044\,\text{kg}$$
$$\text{Net total} = 2.4483\,\text{kg/kg fuel}$$

The stoichiometric mass of air for complete combustion of the fuel is $2.4483 \times 100/23 = 10.645\,\text{kg/kg}$ fuel.

(k)   The actual air supplied by mass is $10.645 \times 1.5 = 15.97\,\text{kg/kg}$ fuel.

(l)   The air fuel ratio by mass is 15.97:1.

Note: The actual air supplied can be expressed in m^3/kg fuel by dividing by air density at 15 °C, thus $15.97/1.234 = 12.94\,\text{m}^3/\text{kg}$ fuel.

Summary to Section 11.7

It is not appropriate to compare the air fuel ratios of the fossil fuels analysed above for a number of reasons. The ratios are quoted by mass and by volume although it would be possible to quote each of them by mass or volume. More importantly, however, the calorific value of each fuel is different – natural gas 30 MJ/kg, light grade fuel oil 40.5 MJ/kg and Medium Rank coal approximately 27.4 MJ/kg. This means that each kilogram of fuel burnt produces a different amount of heat energy. Finally excess air requirements for ensuring complete combustion of the fuel depend on whether the fuel is a gas, liquid or solid.

11.8   Carbon dioxide emissions from fossil fuels

Carbon dioxide emission only occurs from the combustion of the carbon constituent of the fossil fuel. In the cases of methane, fuel oil and coal, there is just one carbon constituent and the combustion of carbon with oxygen to carbon dioxide will therefore use the ratio by mass of 44/12. Natural gas on the other hand has a number of constituents that contain carbon – methane (CH_4), ethane (C_2H_6), propane (C_3H_8), butane (C_4H_{10}). It can also contain the carbon constituents of pentane (C_5H_{12}) and hexane (paraffin (C_6H_{14})). These constituents are known as hydrocarbons and they produce carbon dioxide and water vapour in the combustion process. Table 11.3 lists each of these hydrocarbons along with the equations for complete combustion. The combustion of carbon with oxygen to carbon dioxide in the combustion process requires the use of the appropriate ratios by mass. For example, the molecular mass of the carbon content of methane (CH_4) is 12. The molecular mass of the carbon dioxide produced from the combustion process is 44. The ratio by mass is therefore 44/12.

From Table 11.3, the following ratios by mass for the complete combustion of the carbon content of the hydrocarbons can be identified:

$$
\begin{array}{lll}
\text{Methane} & CH_4 & 44/(1 \times 12) = 44/12 \\
\text{Hexane} & 2C_6H_{14} & 528/(2 \times 12 \times 6) = 528/144 \\
\text{Heptane} & C_7H_{16} & 308/(7 \times 12) = 308/84 \\
\text{Ethane} & 2C_2H_6 & 176/(2 \times 12 \times 2) = 176/48 \\
\text{Butane} & 2C_4H_{10} & 352/(2 \times 12 \times 4) = 352/96 \\
\text{Pentane} & C_5H_{12} & 220/(5 \times 12) = 220/60 \\
\text{Propane} & C_3H_8 & 132/(3 \times 12) = 132/36 \\
\end{array}
$$

Each of these ratios reduce to 44/12. These ratios by mass are used where appropriate for the determination of carbon dioxide emission from burning natural gas. Alternatively the carbon constituents of natural gas can be added together and multiplied by the common ratio 44/12.

Example 11.6
A boiler fired by natural gas has an output of 110 kW and an efficiency of 82%. The analysis by mass of the constituents of the gas is methane 86%, ethane 6.3%, propane 2%, butane 1%, nitrogen 4.2% and carbon dioxide 0.25%. Net calorific value of the gas is $39\,MJ/m^3$.

(a)   From the molecular mass of each constituent determine the density of natural gas and correct to an ambient temperature of 15 °C.
(b)   Determine the emission of carbon dioxide in the products of combustion when the boiler is working at full load. Assume that the air for combustion has negligible CO_2 content.

Solution
(a)   From Section 11.3 the density of natural gas at 0 °C can be calculated from

$$\rho = \text{molecular mass}/22.28,$$

thus

$$\rho = \left(\frac{16}{22.28}\right) \times 0.86 + \left(\frac{30}{22.28}\right) \times 0.063 + \left(\frac{44}{22.28}\right) \times 0.02$$

$$+ \left(\frac{58}{22.28}\right) \times 0.01 + \left(\frac{28}{22.28}\right) \times 0.042 + \left(\frac{44}{22.28}\right) \times 0.0025$$

and therefore at 0 °C, ρ is $0.8256\,kg/m^3$. At ambient temperature (15 °C), the density of natural gas,

$$\rho = 0.8256 \times \frac{273}{288} = 0.7826\,kg/m^3$$

(b)   Efficiency

$$\eta = \frac{\text{output}}{\text{input}} = \frac{\text{output}}{CV \times (VFR)_g}$$

Substituting

$$0.82 = \frac{110}{39\,000 \times (VFR)_g}$$

So

$$(VFR)_g = \frac{110}{39\,000 \times 0.82} = 0.0034396\,m^3/s$$

Calorific value

The reader should note that the *net calorific value* used here and in the following examples using fuel oil and coal in which hydrogen is a constituent, ignores the latent heat in the vapour contained in the products of combustion at atmospheric pressure. A condensing boiler having a heat exchanger in the boiler flue connection is designed to condense this vapour and claim the latent heat, thus improving boiler efficiency. When using a condensing boiler of this type, the *gross calorific value* of the fuel can be used. Converting $(VFR)_g$ to mass flow rate

$$M_g = 0.0034396 \times 0.7826 = 0.002692 \, \text{kg/s}$$

Now finding the mass of carbon dioxide per kg of natural gas

$$CH_4 \ (44/12) \times 0.86 = 3.1533 \, \text{kg of } CO_2$$
$$C_2H_6 \ (176/48) \times 0.063 = 0.231$$
$$C_3H_8 \ (132/36) \times 0.02 = 0.0733$$
$$C_4H_{10} \ (352/96) \times 0.01 = 0.0367$$
$$\text{Sub total} = 3.4943$$
$$CO_2 \ \text{present} = 0.0025$$
$$\text{Total} = 3.4968 \, \text{kg} \, CO_2/\text{kg fuel}$$

Alternatively using the common ratio $44/12$

$$\text{Methane} + \text{Ethane} + \text{Propane} + \text{Butane}$$
$$= 0.86 + 0.063 + 0.02 + 0.01 = 0.953 \, \text{kg carbon}$$

$$\text{Carbon dioxide} = 0.953 \left(\frac{44}{12}\right) + 0.0025$$
$$= 3.4943 + 0.0025$$
$$= 3.4968 \, \text{kg} \, CO_2/\text{kg fuel}$$

$$\text{Carbon dioxide emission} = M_g \times CO_2/\text{kg fuel}$$
$$= 0.002692 \times 3.4968$$
$$= 0.009413 \, \text{kg/s}$$

Thus, carbon dioxide emission at full boiler load is 34 kg/h. This emission of carbon dioxide will be compared with that for burning light fuel oil and Medium Rank coal.

Example 11.7

The analysis by mass of the constituents of light fuel oil are given as 86.2% carbon, 12.4% hydrogen and 1.4% sulphur. The oil is used in a boiler rated at 110 kW and an efficiency of 82%. The net calorific value of the oil is given as 40.5 MJ/kg. Determine the emission of carbon dioxide in the products of combustion when the boiler is working at full load. Assume that the air for combustion has negligible CO_2 content.

Solution

From Table 11.1 the quantity of carbon dioxide produced from burning carbon will be $(44/12) \times 0.862 = 3.1607$ kg/kg fuel.

From $\eta = $ output/input,

$$0.82 = 110/40\,500 \times M_o$$

So

$$M_o = \frac{110}{40\,500 \times 0.82} = 0.0033123 \text{ kg/s} = 11.924 \text{ kg/h}$$

Thus the carbon dioxide emission at full boiler load $= 11.924 \times 3.1607$
$$= 37.7 \text{ kg/h}$$

Example 11.8

The analysis of the constituents of Medium Rank coal by mass is 81.8% carbon, 4.9% hydrogen, 4.4% oxygen, 1.8% nitrogen, 1.9% sulphur and 5.2% mineral matter. The coal is burnt in a boiler rated at 110 kW and an efficiency of 82%. The net calorific value of the coal is quoted as 27.4 MJ/kg. Determine the emission of carbon dioxide in the products of combustion when the boiler is working at full load. Assume that the air for combustion has negligible CO_2 content.

Solution

Carbon dioxide can only be produced from the carbon content of the coal, thus

$$\left(\frac{44}{12}\right) 0.818 = 2.999 \text{ kg/kg fuel}$$

From $\eta = $ output/input,

$$M_c = \frac{110}{27\,400 \times 0.82} = 0.004896 \text{ kg/s} = 17.625 \text{ kg/h}$$

the carbon dioxide emission from the boiler at full load $= M_c \times CO_2/\text{kg fuel}$
$$= 17.625 \times 2.999$$
$$= 52.9 \text{ kg/h}$$

Summary
The foregoing solutions assume complete combustion of the fuels considered. A comparison can now be made of carbon dioxide emissions at full boiler load assuming a common boiler output of 110 kW and a common efficiency of 82%:

Natural gas	34 kg/h
Light fuel oil	37.7 kg/h
Medium Rank coal	52.9 kg/h

The reader can understand why coal use has dwindled at the expense of natural gas in the effort to reduce CO_2 emissions in the UK.

11.9 Annual carbon dioxide emissions from heating plant

The *CIBSE Guide* to current practice uses the term Equivalent Hours of operation at full load EH and EH = 24DD/dt [*CIBSE Guide*, 1986 edition, Section B18.]. This Equivalent Hours at full load formula along with the use of annual Standard Degree Days is the subject of another publication. [*Energy Management and Operating Cost in Buildings*, Spon Press.]

For an office in southern England being occupied intermittently, a typical corrected annual Degree Day total is 900 so for a design indoor to outdoor temperature drop of 23 K,

$$EH = 24 \times \frac{900}{23} = 939$$

For the natural gas fired boiler plant the annual carbon dioxide emission is 34 kg/h × 939 h = 31 926 kg. This is 32 tonnes of CO_2 per year from a plant of just 110 kW! For a well insulated building the plant would serve a treated floor area of about 1600 m², so the annual emission is 31 926/1600 = 19.95 kg CO_2/m². There is also a published *Guide* for energy use in which there are Good Practice and Typical annual emissions of carbon dioxide for four office types [*Energy Consumption Guide* 19, BRECSU, (BRE).], Table *B3* in this *Guide* gives a *CEI* = 4.3 kg/m² for good practice and a *CEI* = 8.3 kg/m² for typical emissions.

Using the carbon to carbon dioxide ratio of 44/12, these carbon emissions translate to CO_2 = 15.8 kg/m² for good practice and CO_2 = 30.4 kg/m² for typical emissions. So the annual emission of 19.95 kg CO_2/m² appears to err toward good practice. However this does not alter the fact that boiler plant of this output produces so much CO_2 annually. The reader should now determine the annual CO_2 emission from the coal fired plant using 939 Equivalent Hours at full load in Example 11.8. The answer is 50 tonnes.

11.10   Development of combustion equations for hydrocarbons

The development of the combustion equations found in Table 11.3 is a relatively easy process. The formation of the equation for Hexane will be used as an example. Hexane is a hydrocarbon and when completely burnt will produce carbon dioxide and water vapour. So

$$C_6H_{14} + O_2 \rightarrow CO_2 + H_2O$$

The combustion of the carbon produces

$$C_6H_{14} + O_2 \rightarrow 6CO_2 + H_2O$$

The combustion of the hydrogen produces

$$C_6H_{14} + O_2 \rightarrow 6CO_2 + 7H_2O$$

Substituting the molecular masses

$$86 + O_2 \rightarrow (6 \times 44) + (7 \times 18)$$

and we have

$$86 + O_2 \rightarrow 264 + 126$$

Rearranging to find O_2

$$O_2 \rightarrow 264 + 126 - 86 \rightarrow 304$$

So the amount of oxygen

$$O_2 = \frac{304}{\text{molecular mass}}$$

$$= \frac{304}{32} = 9.5$$

So the equation for complete combustion of Hexane will be

$$C_6H_{14} + 9.5O_2 \rightarrow 6CO_2 + 7H_2O$$

Decimal places are not usually found in chemical equations, so in this case we multiply through by 2. Thus

$$2C_6H_{14} + 19O_2 \rightarrow 12CO_2 + 14H_2O$$

The reader should now try developing some of the other combustion equations in Table 11.3.

11.11   Chapter closure

Successful completion of this chapter will provide an insight into the principles of combustion of fossil fuels, the calculation of stoichiometric air and the reasons for excess air to ensure complete combustion. The reader will have a knowledge of the development of the combustion equations for hydrocarbons and the procedure for calculating the carbon dioxide emission from boiler plant. An introduction to CO_2 emission bench marking has also been made.

12 Electric heating

Nomenclature

B	overnight charge period (h)
C	specific heat capacity (kJ/kg K)
CA	charge acceptance (kW h)
DDE	daily design energy requirement (kW h)
dt	temperature difference (K)
g	gravitational acceleration at sea level (m/s^2)
h	static head (m)
m	mass (kg)
P	pressure (Pa)
Q	design heat load (kW)
Qg	indoor heat gains (kW)
R	thermal storage factor for the building
r	output controlled by the fan (%)
S	active store at the end of the Charge Acceptance period (kW h)
V	volume (m^3)
v_f	specific volume of water (m^3/kg)
Z	reducing factor accounting for the proportion of Q to be met by the heater the balance being accounted for by indoor heat gains Qg and the thermal storage factor for the building R
η	efficiency (%)
ρ	density (kg/m^3)

12.1 Introduction

As with all forms of heat energy supply there are advantages and disadvantages with their use. Electricity used for space heating and hot water supply is no exception and the first considerations are its cost and carbon emissions in comparison with gas oil and coal. Clearly the reason why there is a substantial difference here is that electricity has to be generated. Coal, oil and gas on the other hand, once having been extracted and prepared for consumer use is available as a fuel. The same is true of nuclear fuel. The production of electricity by renewable means still requires a generator but at least here if natural sources like wind, water or solar are used, electricity generation does not give rise to carbon dioxide emissions or pollution except in the manufacture and maintenance of the generator. One of the disbenefits of electricity only generation from fossil fuels is the low production efficiency that at best is around 40% and commonly nearer 30%. This means that for each kilogram of fossil fuel 70% of it is wasted mostly in the cooling towers. The drive toward co generation or combined heat and power plants is seen as one way to increase the overall efficiency to something more acceptable.

Electricity generation in the UK has moved forward since the Kyoto Summit. As a result of changes in generation from coal to gas and some renewables together with the upturn in combined heat and power, its emission factor for carbon dioxide has fallen from $0.7\,kg\,CO_2/kW\,h$ to $0.5\,kg\,CO_2/kW\,h$. This compares with an emission factor for heating by natural gas of $0.2\,kg\,CO_2/kW\,h$. If renewables relating to electricity generation receive further substantial Government support, the carbon dioxide emission factor for electricity generation will fall below $0.5\,kgCO_2/kW\,h$.

Another view that has mounting credence in respect of CO_2 emission reduction associated with buildings is the concept of life cycle costing which considers financial, social and environmental factors as opposed to capital cost alone. Evidence shows that life cycle costs of a building in addition to its initial capital cost need to be part of the 'cost' equation.

An enlightened approach to procurement will consider the energy consumption and CO_2 emissions of products and materials used in the construction process and during the life cycle of the building. In addition, financial appraisal will include capital cost and life cycle costs of maintenance and replacement. However, notwithstanding the energy and emission factors that mitigate against the use of electricity in buildings it is unlikely that it can be replaced.

Electrothermal storage systems are a means of heating which make use of off-peak electricity. There are two advantages to be gained here. First, there is the lower tariff available during off-peak hours and second, it assists the power generators by levelling out the peaks and troughs in electricity demand from the grid. The charge acceptance period (off-peak period) is approximately seven in every twenty-four-hours day and occurs usually from 00.00 to 07.00 h. There are a number of methods of thermal storage available:

(i)   static solid core storage heaters with element ratings from 1.7 to 3.4 kW;
(ii) fan assisted solid core storage heaters having element ratings of between 4 kW and 7.5 kW;
(iii) fan assisted solid core storage heaters suitable for centralized ducted warm air or as dry core boilers with air to water heat exchangers and circulating pump serving low temperature hot water radiators;
(iv) electro thermal water storage employing immersion heaters or the electrode boiler. The electrode boiler is also used for swimming pool heating.

Solid core temperatures can reach 900 °C by the end of the charge accept-ance period. Final storage temperature is limited to the type of storage material used. This chapter will discuss the use and application of these methods of heating except for item (iii) which is not now in common use.

12.2   Static storage heaters

There are three sizes of static storage heater as shown in Table 12.1. Each is fitted with input and output controls. Weather compensated control is also available for these heaters. Input control is achieved by thermostatically con-trolling the core temperature during the charge acceptance period (CA). Output control is done by varying the air flow over the core material using a damper. Both these controls are manually set. A further input temperature control is available and effectively responds in a similar way to weather compensated control used in LTHW systems. The instantaneous output from a static storage heater fitted with a 3.4 kW element and having a 7 h CA period will be:

$$\text{Instantaneous output} = 7 \times \frac{3.4}{24} = 1\,\text{kW}$$

The heater is of course functional for 24 h/day. Similarly the instantaneous output from a static storage heater with a 2.55 kW element on a CA of 7 h is $7 \times 2.55/24 = 0.74\,\text{kW}$.

The heater output however is not constant over time and its maximum output will occur at the end of the charge acceptance period, usually about 07.00 h. Its minimum output will occur at about midnight at the commence-ment of the CA period. Using the instantaneous output would allow the determination of the number of heaters required for a given design heat loss. However since this type of heating is continuous it does not account for the heat stored in the building structure over time. When a building is continu-ously heated 24 h/day, 7 days/week over the entire heating season, the effect on indoor temperature resulting from fluctuations in outdoor temperature depends upon the thermal storage capacity of the building. Heavyweight buildings have a greater damping effect than lightweight buildings and the effect on indoor temperature resulting from swings in outdoor temperature therefore has less effect in a heavyweight building than it does in a lightweight

Table 12.1   Static storage heaters

Static storage heater	Element rating kW	Night time charge acceptance (CA) kW h
Small	1.7	12
Medium	2.55	18
Large	3.4	24

CA based on a 7 h charge period

Table 12.2   Classification of buildings

Class of building by mass	Thermal response factor (intermittent heating) f_r	Thermal storage factor (continuous heating) R
Heavy Buildings with appreciable areas of solid walls and partitioning	6.0	0.25
Medium Single storey buildings as for Heavy	4.0	0.17
Light Single storey buildings of factory type construction with no solid partitions	2.5	0.11

structure. For continuous heating this is accounted for by applying the thermal storage factor (R) in the calculation of Z Table 12.2.

Indoor heat gains (Qg) can also be accounted for in determining the design heat load (Q) and the mean value of Qg is taken over the twenty-four hours-day. Thus reducing factor $Z = 1 - (Qg/Q) - R$ and the Daily Design Energy requirement, $DDE = 24QZ\,kW\,h$. The design procedure requires that a heat balance is drawn such that $DDE = CA$. Table 12.1 gives values of R. It must not be confused with the f_r which is used to determine the overload capacity for heating plant serving space heating systems that operate intermittently. It is included in Table 12.2 since building classification is associated with the calculation of the f_r.

Example 12.1
A room located in a building classified as medium weight measures 12 m by 8 m with one longer wall exposed and has a design heat load of 6 kW. There are three windows symmetrically located on the wall. Heat gains are estimated as 17 W/m² during 10 h of use. Determine the DDE requirement and hence evaluate the size and number of static storage heaters required to offset the design heat load.

Solution
Although it does not form part of the solution the rate of heat loss per square metre of floor is $6000/12 \times 8 = 62.5\,W/m^2$ which indicates a high level of thermal insulation.

Mean heat gains $= 17 \times 10/24 = 7\,\text{W/m}^2$ and $Qg = 7 \times 12 \times 8 = 672\,\text{W}$

$$\frac{Qg}{Q} = \frac{0.672}{6} = 0.112$$

and from Table 12.2, $R = 0.17$. Thus

$$Z = 1 - 0.112 - 0.17 = 0.72$$

and

$$\text{DDE} = 24QZ = 24 \times 6 \times 0.72 = 104\,\text{kW h}$$

From Table 12.1, three storage heaters rated at 3.4 kW

$$\text{CA} = 3 \times 3.4 \times 7 = 71.4\,\text{kW h}$$

This is below the Daily Design Energy (DDE) by $103.7 - 71.4 = 32.3\,\text{kW h}$. If two direct heaters are also employed and the occupation period is 10 h the output of each heater is $32.3/10 \times 2 = 1.614\,\text{kW}$. These could be conveniently located on the wall opposite to the external wall. Alternatively two more static heaters could be considered. From Table 12.2 two static heaters rated at 2.55 kW give

$$\text{CA} = 2 \times 2.55 \times 7 = 35.7\,\text{kW h}$$

Thus the total charge acceptance

$$\text{CA} = 71.4 + 35.7 = 107.1\,\text{kW h}$$

This compares with DDE $= 104\,\text{kW h}$, and this equates well with the CA requirement and the heat balance CA $=$ DDE is preserved.

Summary

Either 3 static heaters rated at 3.4 kW and 2 direct heaters rated at 1.614 kW *or* 3 static heaters rated at 3.4 kW and 2 static heaters rated at 2.55 kW.

The reader will note that instantaneous output from the five static storage heaters amounts to 4.48 kW whereas the net heat loss is 4.37 kW.

12.3   Fan assisted storage heaters

There are four standard sizes of fan assisted storage heaters as shown in Table 12.3. The Charge Acceptance (CA) figures quoted assume a 7 h charge period. Since the heat storage material by its nature is dense the floor loadings in

Table 12.3 Fan assisted storage heaters

Element rating kW	Night time charge acceptance (CA) kW h	Approximate size mm	Approximate floor loading kN/m²
4.0	28	1050 × 670 × 255	7.14
5.0	35	1250 × 670 × 255	7.40
6.0	42	1250 × 670 × 255	7.40
7.5	52.5	1250 × 670 × 255	8.93

CA based on a 7 h charge period

kN/m^2 should be noted as they may exceed standard design floor loading in the building.

The active store

$$S = (24QZ - BQZ) \quad \text{kW h}$$

and

$$\text{DDE} = \frac{S \times 100}{r} \quad \text{kW h}$$

The design procedure again requires that a heat balance is maintained thus

$$\text{DDE} = \text{CA}$$

Input control is achieved by thermostatically controlling the core temperature during the charge acceptance period. A damper housed within the unit is used to allow the mix of the room air with the hot air from the heater core to a maximum temperature of 65 °C. The heater output fan is controlled by a room thermostat and when it is off the storage heater output is reduced significantly – details are given by the manufacturer.

Example 12.2

An office having a floor area of 15 m × 10 m has a design heat load of 12 kW. The estimated heat gains amount to 22 W/m² of floor surface during eleven hours of use. Given that the heater output fan controls 80% of the total energy stored in the heater, determine the size and number of fan assisted storage heaters to offset the design heat load. The office is located in a building classified as a heavyweight.

Solution

$$\text{Mean heat gain} = \frac{22 \times 11}{24} = 10 \, \text{W/m}^2$$

$$Qg = 10 \times 15 \times 10 = 1500 \, \text{W}$$

$$\frac{Qg}{Q} = \frac{1.5}{12} = 0.125$$

From Table 12.2, $R = 0.25$

$$\text{Reducing factor } Z = 1 - 0.125 - 0.25 = 0.625$$

Active heat store at the end of the CA period

$$S = 24QZ - BQZ = (24 \times 12 \times 0.625) - 7 \times 12 \times 0.625 = 127.5\,\text{kW h}$$

$$\text{DDE} = S \times \frac{100}{r} = 127.5 \times \frac{100}{80} = 160\,\text{kW h}$$

From Table 12.3 three fan assisted storage heaters rated at 7.5 kW will give a Charge Acceptance

$$\text{CA} = 3 \times 7.5 \times 7 = 157.6\,\text{kW h}$$

This equates with the daily design energy requirement of DDE = 160 kW h.

Summary
The heater output fan is controlled by a room thermostat but it is assumed that it is off during the charge acceptance period. A 7.5 kW rated fan assisted storage heater has an instantaneous output of $7.5 \times (7/(24 - 7)) = 3.1\,\text{kW}$ at the end of the charge acceptance period. So the instantaneous heater output $3 \times 3.1 = 9.3\,\text{kW}$ whereas the net heat loss is 8.7 kW.

12.4   Electro thermal storage

This type of heating is normally achieved using the electrode boiler of which there are two types: The low voltage boiler rated at 400–600 V ac and the high voltage boiler rated at 11 kV ac. A diagram of the electrode boiler is shown in Figure 12.1.

Water is heated by passing a current directly through the water between two electrodes. The electrical load and hence the heat input to the water is varied by increasing or decreasing the length of the path that the current has to take by employing electrode shields or sleeves. The electrodes dissipate considerable thermal energy requiring rapid water circulation for which an integral circulating pump is used. Since water resistance to electrical current decreases with rise in temperature a complex shield positioning mechanism is required. The water that is heated within the boiler during the charge acceptance period is transferred to a thermal storage vessel via the loading circuit shown in Figure 12.2. At the end of a design day, the water in the thermal storage vessel will be at a final uniform design temperature. At the

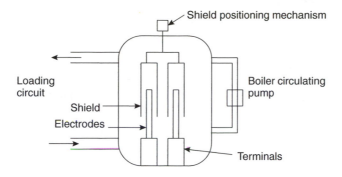

Figure 12.1   Diagram of an electrode boiler.

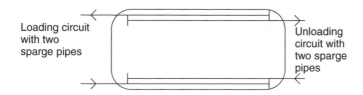

Figure 12.2   Thermal storage vessel with sparge pipes to assist in stratification of the water contents.

end of the charge period the vessel water will be at the initial uniform design temperature. At each point there must be little or no temperature gradient within the vessel. To ensure that this is the case sparge pipes are located in the vessel on the loading and unloading circuits (see Figure 12.2).

Figure 12.3 shows a typical layout of an electro thermal storage system. The unloading circuit may in fact consist of a number of circuits, for example, one or more zones of radiators, circuits to air heater batteries, fan coil circuits, primary hot water service circuits etc. Note the use of the antisyphon pipe which prevents high temperature water from backing up the feed and expansion pipe to the header tank. It should be sized to contain the volume of expansion water generated during the charge period. The storage temperature at the end of the charge period (initial storage temperature) must have a minimum anti-flash margin of 10 K between it and its boiling point which in turn is dependent upon the static pressure exerted by the height of the header tank above the plant. The thermal storage vessels must be efficiently insulated to ensure minimum heat loss. Design procedures for the electro thermal storage plant include determination of generator output, power supply requirement, thermal storage capacity and volume of expansion water resulting from the rise in water temperature attributed to the electrode boiler during the charge acceptance period. This is required to size the antisyphon pipe and header tank.

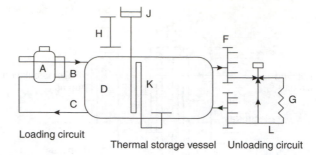

Loading circuit   Thermal storage vessel   Unloading circuit

Figure 12.3 Electrothermal storage plant. *Notation*: A, electrode boiler; B, boiler circulating pump; C, primary pump on loading circuit; D, thermal storage vessel; F, flow and return headers on the unloading circuit; G, weather compensated control on a typical radiator zone; H, static head on the system; J, feed and expansion or header tank mains with water service connection omitted; K, antisyphon feed and expansion pipe; L, circulating pump on radiator circuit.

Example 12.3
A system of electro thermal storage is being considered for a building that has a gross heat load of 50 kW. The idling losses which occur during the hours when the building is unoccupied are estimated at 10% of the gross heat load. From the data find the generator output, electrical power requirement, thermal storage capacity and expansion volume at the end of the charge period.

Data
Available system static head 15 m, charge period 7 h, antiflash margin 10 K, temperature rise in thermal storage vessel during charge period 30 K, electrical power supply 400 V 3phase, occupancy 08.00 h–18.00 h, efficiency of the electrode boiler 96%.

Solution

$$\text{Idling losses} = 50 \times \frac{10}{100} = 5\,\text{kW}$$

$$\text{Occupancy period} = 10\,\text{h}, \ \text{idling period} = 14\,\text{h}$$

$$\text{Generator output} = (50 \times 10) + (5 \times 14) = 570\,\text{kW h}$$

For a 7 h charge period generator output is $570/7 = 81.4\,\text{kW}$.
Electrical power requirement in amperes

$$I = \frac{W}{V \times \sqrt{3} \times \eta} = \frac{81.4 \times 1000}{400 \times \sqrt{3} \times 0.96} = 123\,\text{amps}$$

Thermal storage capacity in kilograms of stored water:

$$m = \frac{\text{energy at the end of the charge period in kJ}}{C \times dt}$$

$$m = (570 - (5 \times 7)) \times \frac{3600}{4.2 \times 30} = 15\,286\,\text{kg}$$

gauge pressure resulting from a static head of 15 m

$$P = h \times \rho \times g \quad \text{Pa}$$
$$P = 15 \times 1000 \times 9.81 = 147\,150\,\text{Pa}$$

$$\text{Absolute pressure} = \text{gauge pressure} + \text{atmospheric pressure}$$
$$= 147\,150 + 101\,325$$
$$= 248\,475\,\text{Pa} = 2.48\,\text{bar}$$

from the *Thermodynamic and Transport Properties of Fluids* [by Rogers and Mayhew], saturation temperature $= 127\,°C$ and storage temperature $= 127 - 10\,K$ antiflash margin $= 117\,°C$. This is the initial temperature of the stored water at the end of the charge period. From the same tables specific volume of water v_f at $117\,°C$ is $0.001\,058\,\text{m}^3/\text{kg}$. The thermal storage volume

$$V = m \times v_f = 15\,286 \times 0.001058 = 16.17\,\text{m}^3$$

the expansion volume of water resulting from an initial storage temperature of $117\,°C$ and a final storage temperature of $117 - 30 = 87\,°C$ can be calculated from

$$\text{expansion volume} = \frac{V(\rho_1 - \rho_2)}{\rho_2} = \frac{16.17(967 - 945)}{945} = 0.376\,\text{m}^3$$
$$= 376\,\text{litres}$$

The water densities at the initial ρ_1 and final ρ_2 storage temperatures are obtained from the tables of properties of fluids used above where $\rho = 1/V_f$. Note the final storage temperature at the end of the design day is $87\,°C$.

Summary
Generator output $= 81.4\,\text{kW}$, electrical power required $= 123\,\text{amps}$, thermal storage capacity $= 16.17\,\text{m}^3$, expansion volume for sizing the antisyphon pipe and the header tank $= 376\,\text{litres}$.

12.5   Swimming pool heating

If an indoor swimming pool is to be heated, the dry bulb temperature of the pool hall should be in excess of the pool water temperature to inhibit condensation occurrence on the inner surfaces of the pool hall. When the pool is up to temperature and uncovered for use there will be a latent heat loss from the pool and a corresponding latent heat gain to the pool hall. This requires the hall to be mechanically ventilated to ensure against surface condensation. The latent heat gain should be reclaimed in the air handling plant by using it to raise the temperature of the incoming fresh air during the winter season in order to save on fuel costs and in the interests of energy conservation. When the pool is in use, the latent heat gain to the pool hall is increased as a result of disturbance at the water surface by the pool users. There is also a latent heat gain from the users of the pool.

The total heat loss from surface of a sunken swimming pool is given by

$$\sum q = q_e + q_c + q_r + q_{cd} \quad \text{W/m}^2$$

The equations for latent heat loss (q_e) from the still surface of the pool and the heat convection gain (q_c) to the pool surface are empirical and therefore approximate. The equations for heat radiation loss (q_r) from the pool water surface and conduction loss (q_{cd}) from the pool tank are rational formulae. The following formulae are taken from the *CIBSE Guide* Section C.

$$q_e = (91.5 + 77.6u)(P_s - P_w)$$

$$q_c = 3.18(u)0.8(t_w - t_a)$$

$$q_r = 5.67 \times 10^{-8} \times e(T_w^4 - T_r^4)$$

$$q_{cd} = U(t_w - t_g)$$

The example that follows, identifies the notation and applies these equations to an indoor swimming pool.

Example 12.4

From the data relating to an indoor swimming pool, determine:
  (i)   net heat loss from the pool water surface;
  (ii)   heat loss from the pool tank;
  (iii) conventional boiler output without the pool cover to maintain the water temperature;
  (iv)   conventional boiler output with the pool cover in place to maintain the water temperature;
  (v)   net latent heat gain to the pool hall;
  (vi)   daily design energy requirement with and without the pool cover in place to maintain the water temperature;
  (vii) output from an electrode boiler to maintain the water temperature assuming regular use of the pool cover.

Data

Pool size $15 \times 7 \times 1.7$ m mean depth, pool use 12 h/day, water temperature t_w 27 °C, hall air temperature t_a 30 °C, relative humidity 71%, mean radiant temperature in the pool hall t_r 26 °C, air velocity over the pool surface u 0.5 m/s, emissivity of water e 0.96, mean U value of pool tank 3.0 W/m² K, ground temperature t_g 10 °C, charge acceptance period CA 7 h.

Solution

Note that T_w and T_r are absolute temperatures of the water surface and the mean radiant temperature of the pool hall. From the tables of properties of humid air in the *CIBSE Guide Section C* or [*CIBSE Concise Handbook*] saturation vapour pressure P_s at 30 °C db is 4.242 kPa and the vapour pressure P_w at 71% relative humidity is 3.008 kPa. Substituting the data into the equations we have:

$$q_e = 161 \text{ W/m}^2$$

$$q_c = -5.48 \text{ W/m}^2$$

$$q_r = -5.85 \text{ W/m}^2$$

$$q_{cd} = 51 \text{ W/m}^2$$

You should now confirm these solutions

(i) The net heat loss from the pool water surface is $(161 - 5.48 - 5.85)$ $(15 \times 7) = 15\,715$ W.

(ii) The heat loss from the pool tank is $51((44 \times 1.7) + (15 \times 7)) = 9170$ W.

(iii) Net output required from a conventional boiler without the pool cover is $15\,715 + 9170 = 24\,885$ W.

(iv) If a pool cover is used the latent heat loss from the pool is cancelled leaving a net heat gain to the pool of $(5.48 + 5.85)(15 \times 7) = 1190$ W. So the net output required from a conventional boiler to maintain the water temperature will be $(9170 - 1190) = 7980$ W.

(v) The latent heat gain to the pool hall is $161 \times (15 \times 7) = 16\,905$ W.

(vi) The DDE without the pool cover in place is $24\,885 \times 24 = 597$ kW h the DDE with the cover in place for 12 h is $(24\,885 \times 12) + (7980 \times 12) = 394$ kW h.

(vii) Net output of an electrode boiler having a charge acceptance of 7 h is $394/7 = 56.3$ kW.

Summary

In practice, as the pool tank is sunk into the ground the mass of earth surrounding the tank warms up over time and acts as a thermal stabilizer, so boiler output in (iv) is reduced. To reduce the time it will take to warm the water from cold, the pool cover should be in place continuously during the warm up period. The final size of the boiler may depend upon the maximum length of time that can be allowed to get the pool up to its operating temperature. For example, assuming a conventional boiler of 25 kW output which is equivalent to solution (iii) and needed to maintain the pool at 27 °C when it is in use, the approximate time to raise the pool contents by one degree can be calculated as

$$\text{Mass of water} = 15 \times 7 \times 1.7 \times 1000 = 178\,500\,\text{kg}$$

and since

$$\text{power} = \frac{\text{mass} \times \text{specific heat} \times \text{d}t}{\text{time}}$$

$$\text{time} = \frac{178\,500 \times 4.2 \times 1}{25} = 29\,988\,\text{s} = 8.33\,\text{h}$$

If the initial temperature of the water is 10 °C the approximate time for it to reach 27 °C will be $(27 - 10) \times 8.33 = 142\,\text{h}$. This is equivalent to 6 days. In fact it will be longer than this as heat loss will occur during the heat up period.

12.6   Further reading

CIBSE Guide B1 Heating
Building Services OPUS Design File

12.7   Chapter closure

You now have the skills to design a variety of storage heating systems using off-peak electricity. It is important however to investigate the market to familiarize yourself with the products that are available. Element ratings of static and fan assisted storage heaters for example vary from manufacturer to manufacturer. This of course affects the maximum charge acceptance for the heater.

District and community heating 13

Nomenclature

CHP	combined heat and power
d	pipe bore (m)
E	modulus of elasticity (Pa)
f	stress (Pa)
F	force (N)
HTHW	high temperature hot water
HWS	hot water service
I	moment of inertia $(cm)^4$
L	offset (m)
LTHW	low temperature hot water
MTHW	medium temperature hot water
w	deflection (m)
z	distance in metres of outside of pipe from the neutral point $= d/2$

13.1   Introduction

'District' heating in the UK is traditionally divided into two types namely that serving flats, apartments and housing, for example under common or private ownership and that serving a variety of building owners that can include commercial, retail, industrial, residential and local authority. The former is normally called community heating and the latter district heating. District heating requires considerable capital investment and is normally undertaken by organizations that specialize in this area of work. Other areas of specialism associated with district heating include the pipework used for the external mains, heat metering equipment, community waste and medical waste incineration plant and combined heat and power plant.

Both district heating and community heating in recent years have been associated with power generation in the form of combined heat and power (CHP) where the electrical power generated is used locally and any excess power supplied to the national grid. In one community heating system using CHP blocks of flats are heated from a central plant and the lifts, entrances, corridors and landings are powered and lit from the electricity generated. As part of its district heating system Southampton City has a geothermal heating installation in which heat energy is imparted to water from hot rocks 2 km below the surface. If power is accounted for in pumping activities there is otherwise no primary fuel used and hence no pollution or CO_2 emissions. Waste incineration is used to generate heat and power in district heating as also is biomass. Nottingham city has a community waste incineration system as part of its district heating scheme. After combustion the residue is separated into ferrous/non ferrous material and the residue remaining is compacted and sold as fertiliser.

This chapter focuses on district heating which by its nature is more complex than community heating. However, with the use of CHP community heating schemes for residential accommodation, such as blocks of flats and apartments are in common use in the UK. Much of the material included here is applicable to community heating.

13.2   Factors for consideration

The feasibility study for a new scheme will need to investigate and research a number of areas before an initial decision can be reached. The following will give an insight to some of the considerations.

TYPE AND CONCENTRATION OF CONSUMERS

Broadly speaking, the greater the variety the better and could include commercial, industrial, retail, residential – housing association/private, local authority – schools, libraries, museums, art galleries, leisure centres.

BASE LOAD CONSUMER

A considerable advantage is brought to bare in a feasibility study if a base load consumer is available. This would provide a fairly constant and substantial requirement for heating throughout the year ensuring continuous use of plant. Another base load usually available is that of the need for hot water supply particularly in residential sector.

CONSUMER ACCEPTANCE

One of the surveys required in a feasibility study is that allowing an analysis of consumer acceptance to the concept of being connected to a heating utility. A major factor is the charge for the service which needs to be at least 10% below the cost for running and maintaining individual boiler plant. Another factor would be the security of heat supply.

WASTE DISPOSAL

If waste accrues as a result of generating heat/power as it would do in waste incineration, its disposal needs careful consideration.

ENVIRONMENTAL FACTORS

Ideally the heat and power generating plant needs to be located near, if not in the district benefiting from the scheme. However, this mitigates against the noise disturbance and pollution which may result. Waste incineration plant needs feeding which implies a fairly continuous supply of refuse preferably via rail but otherwise via dedicated road network. Fossil fuel may have to be carried in bulk by road/rail. Pollution on the other hand should be well controlled with the use of sophisticated scrubbing and cleaning of the combustion products and the aid of a tall chimney to disperse the products. The environmental and social impact is considerable not only from the generating plant but also from laying the distribution mains. Finally there is the question of sustainability. Does the impact on the local area allow a sustainable future for the town or city.

ENERGY UTILIZATION

A district heating proposal should show a clear advantage in its use of energy from fossil fuel, waste and renewable sources over local heat generation. It therefore should favour the use of waste and renewable energy together with low grade fossil fuel which is difficult to use in relatively small local plant. Residual fuel oil for example is cheaper than light grade oil but it requires preheating to transport it by road and more importantly on site. At ambient conditions it will not flow because of its high viscosity. Currently, however, low grade fossil fuel is not necessarily the only choice in district heating schemes with CHP. The gas turbine of course requires natural gas. Another reason for this change of emphasis is the tighter control imposed on pollutants discharging into the atmosphere from the combustion of low grade fossil fuel such as brown coal and residual heavy grade fuel oil.

MAINTENANCE PHILOSOPHY

One of the lessons learnt in early district heating schemes is the importance of getting the maintenance of plant and external pipework right. A sure way to achieve this is to engage the contractors responsible for the installation of the distribution mains, for example, in a maintenance contract, following completion. In other words, the contract should include both installation and maintenance for an agreed period of at least 20 years. There is always the possibility that a contractor may go out of business of course, but this is a risk that should be included in the risk assessment in the feasibility study.

TOPOGRAPHY

The contour lines of the locality in which the scheme is going need consideration since levels are important in relation to pumping water in plant and distribution pipework not least within the consumers premises. High rise apartments for example may require special attention.

LIKELY INVESTORS

The organization that specializes in district heating will have access to potential investors. The local authority will have an interest in the scheme since it will impinge upon the road network of the locality for which it is responsible.

RISK ASSESSMENT

Risk assessment will give an overview of potential disasters and assist in developing a system of checks and balances that will provide a proactive methodology for either insuring against disaster or managing disruption when it happens. A list of areas that need inclusion in a risk assessment might include breakdown of plant, distribution mains, controls, IT facility, financial management; contractor bankruptcy; strikes; non availability of replacement plant and flooding.

13.3   Choice of system and operating parameters

ENERGY SOURCES AVAILABLE

These would include the most likely energy sources for the scheme – fossil fuel available, type of waste and type of renewable energy. Access to these sources is also an issue as it is likely to involve dedicated transport and road/ rail infrastructure.

HEAT SUPPLY RATIOS

Assessment of approximate ratios of process, heating, hot water supply loads and power loads are needed for the feasibility study. This will have a direct bearing on the choice of plant.

HEATING MEDIUM

The heating medium needs consideration. It is invariably water because of its high specific heat capacity and may be LTHW, MTHW or HTHW. Steam might be considered at the central plant because of its high latent heat.

HEAT EXCHANGE

If high temperatures are employed from the central plant it will be necessary to reduce the temperature using heat exchangers at substations local to residential consumers for example. Industrial consumers may want high temperature distribution for their process and space heating.

CONSUMER SUPPLY

The distribution pipework layout should seek to ensure a ring main arrangement, so that, if a consumer's supply fails, a temporary reconnection can be made quickly. Mains failure invariably occurs in the winter months and temporary connections are sometimes made with the view to restoring the original connection in the following summer.

BASIS FOR CHARGING

There are three ways in which the consumer can be charged for energy use:

1. Flat rate, which avoids the use of metering equipment and hence its initial cost and maintenance;
2. Service charge plus charge for energy actually used;
3. Charge for energy use that includes the service charge.

One of the causes for consumer concern where metering was used has been in the accuracy of the metering equipment and in the recording process.

FUTURE REQUIREMENTS

There will be a need to allow for future extensions to the scheme, so, it is important to know the localities that might see expansion. These will be available from the Local Authority's strategic plan.

DIVERSITY

The feasibility study will address the diversity factor that will be applied to the plant for sizing purposes. Clearly the central plant is not sized on the total net load when all consumers are on line. As a general rule, the lower the number and diversity of consumers connected to the scheme the higher will be the diversity factor. The CIBSE Guide book A recommends a diversity factor of 0.7 for district heating schemes that have a wide spread and high numbers of consumers.

13.4   Heat distribution pipes [*Distribution pipework, Perma – Pipe Services Ltd.*]

There are four different systems of underground district heating pipe networks in use worldwide.

SINGLE PIPE SYSTEM

This system (Figure 13.1) is widely adopted in the Soviet Union to transport HPHW or steam over long distances. Consumers use water or steam for heating buildings and use in swimming pools, the cool water or condensate at the far end of the scheme finally going to waste.

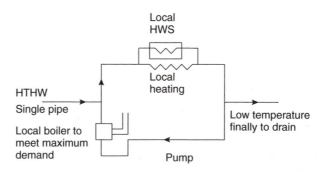

Figure 13.1   The single pipe system.

TWO-PIPE SYSTEM

This arrangement (Figure 13.2) is common in Europe with hot water pumped direct to the consumer serving radiators and an indirect hot water cylinder, for example, or if it is high temperature, indirectly via a heat exchange substation.

THREE-PIPE SYSTEM

While being more expensive to install and maintain, the three-pipe system (Figure 13.3) does offer an advantage in terms of economy of heat supply. In addition to the main flow and return pipes that constitute the two-pipe system this arrangement includes a smaller diameter third flow pipe which is used in the summer for the base load. The common return serves both supplies. The advantage of this system over the two-pipe arrangement is that while in the winter the hot water is pumped out of the central plant through both flow pipes, in the summer the larger diameter flow is closed down and the heat is supplied via the smaller pipe. In this way the network's summer heat losses are reduced. There is also some flexibility here in that with appropriate interconnections a mains failure in either of the flow pipes during winter allows the remaining live main to provide a degree of back up.

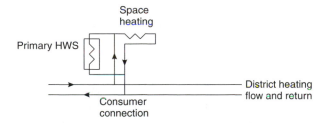

Figure 13.2   Two-pipe distribution with direct space heating.

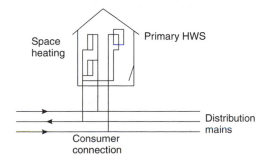

Figure 13.3   Three-pipe distribution with direct space heating.

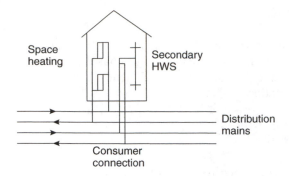

Figure 13.4   Four-pipe distribution with direct space heating.

FOUR-PIPE SYSTEM

In this system (Figure 13.4) there are two flow and two return pipes. One pair is used for transporting domestic hot water and the other, hot water for space heating. This system is not commonly adopted for two reasons, namely, resulting from cost and also because the transport of domestic hot water will incur the probability of scale and corrosion as the mains are handling raw water. Centralized water treatment plant incurs further capital and maintenance costs.

MANUFACTURE AND INSTALLATION OF HEAT DISTRIBUTION PIPES

A study of 15 district heating schemes in the UK undertaken by the Building Research Establishment some years ago found that the heat losses from the distribution mains commonly exceeded 25% of the annual energy consumption. This puts into context the importance of minimizing the heat losses from underground distribution pipes even if it otherwise helps to de ice the roads and pavements under which they are located.

There are specialist manufacturers of pipe systems for district heating mains. The pipe systems available are listed below.

THE INSULATED BONDED PIPE SYSTEM

This system is factory prefabricated and comes pre-insulated. The pipework is welded together on site and the joints heat sealed with an outer sleeve. Bends, tees, valves, vent and drain points are connected to pipe tails and pre-insulated in the factory. Anchors also are fitted to pipe tails and pre-insulated. The pipes are buried and because they cannot be visually inspected the system is fitted with monitoring cables that can detect the presence of moisture. The purpose of the monitor is to allow the detection location and repair of the defect before serious damage occurs. The principle of operation is by measuring the resistance of a loop and locating faults by time domain reflectometry.

DOUBLE PIPE SECONDARY CONTAINMENT SYSTEMS

Containment piping systems are designed with an air space between the heat carrying conduit and the outer or secondary containment pipe. The air space is sufficient in volume to allow initial containment of a leak and also eventual draining and drying out of the system. In addition, the space is large enough to accommodate the internal pipe supports, thermal expansion and leak monitoring devices. Different materials can be used for the heat carrying conduits and secondary containment pipe according to the location of the system and the medium to be transported. One, two, three or four heat carrying conduits can be located within the containment pipe. Materials in use are polypropylene, PVC, polythene, GRP, stainless steel, and coated carbon steel. Fittings are complete with tails and factory made to order.

INSULATED PIPE IN PIPE

This system is used for transporting HTHW or steam and consists of the heating medium conduit thermally insulated with an outer protection sleeve housed in a carrier pipe. The carrier pipe houses insulated supports to guide the heating medium conduit within the carrier pipe. The thermal insulation used is either calcium silicate or high density mineral wool. The system is prefabricated including fittings and anchors. Valves, axial compensators, drains, vents are housed in prefabricated steel-manhole-containers for burial in the ground. When the system is buried directly in the ground cathodic protection is required. Corrosion of the conduit is the result of an electrochemical reaction between the metal pipe and the soil. A galvanic potential exists between the anodic and cathodic areas of the pipe and this drives a current through the soil. Cathodic protection is a process that makes the entire pipe system a cathode that does not corrode, by means of a sacrificial anode which does. The sacrificial anode must therefore have enough material (usually magnesium) to last for a given time so that replacement maintenance can be programmed. In order to maintain integrity of the cathodic protection circuit, the service pipes have to be electrically insulated from above the ground pipework.

13.5   Heat and power generating plant

There are two philosophies related to heat and power generation, namely which is the main requirement – heat or power. The focus for district and community heating schemes is clearly heating, so, the heat load for the scheme will be the major component.

Power only generation from fossil fuel suffers from very low overall efficiencies and hence waste of the primary fuel as well as excessive emissions of pollutants into the atmosphere. The use of wind turbines on the other hand that generate electricity produce no pollutants although overall conversion efficiency is also low.

There is therefore some pressure on the electricity generating companies to consider CHP or renewable energy when a power only station is decommissioned. The effect on overall efficiency when the CHP option is used is to raise it from 30% to as high as 75%. It is necessary however for the generator to find a customer for the heat generated from the CHP plant. Community and district heating can also benefit from the application of CHP as electricity can be used to power the plant with the surplus sold to the national grid. However, the main focus here is in generating heat. In this context combined heat and power may therefore be considered as part of the generating equipment for district heating and there are a number of possible scenarios. Figures 13.5, 13.6 and 13.7 are line diagrams showing CHP generated from the gas turbine using a waste heat boiler, the steam turbine using a refuse incineration plant and the diesel engine using a waste heat boiler. You should note that the boiler plant shown in Figure 13.5 can be fired from residual fuel oil whereas the gas turbine of course requires natural gas. Waste heat fire tube boilers in the gas turbine and diesel engine plants are connected to the high temperature exhaust gases and produce steam or HTHW. Conventional boiler plant would be interconnected with the CHP plant.

The advantage with the marine diesel engine is that it can burn residual oil and three further heat exchangers take heat away from the turbo-charged air cooler, engine water jacket and oil cooler, thus producing a better overall plant efficiency. It is also possible to employ an economizer in the flue gases from the waste heat boilers as long as flue gas dew point is not reached. This will add a further heat exchanger to extract heat from the generating plant. Because of the high pressures and temperatures produced, waste incineration plant is also likely to be used to generate electricity as well as heat. There are

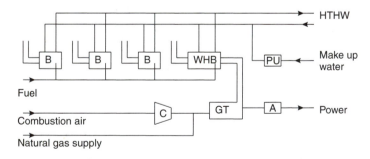

Figure 13.5 CHP district heating using a gas turbine. *Nomenclature*: B, boiler; WHB, waste heat boiler; GT, gas turbine; C, air compressor; PU, pressurization unit; A, alternator; HTHW, high temperature hot water.

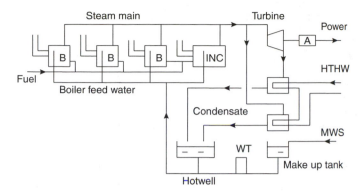

Figure 13.6   CHP district heating using waste heat incinerator and stream turbine. *Nomenclature*: B, boiler; INC, refuse incinerator; A, alternator; HTHW, high temperature hot water; MWS, mains water service; WT water treatment plant. The HTHW return passes through two steam to water heat exchangers. Pressurization unit omitted.

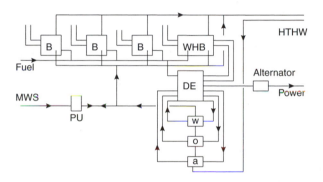

Figure 13.7   CHP district heating using the diesel engine. *Nomenclature*: B, boiler; WHB, waste heat boiler; HTHW, high temperature hot water; DE, diesel engine; w, engine water jacket cooler; o, engine oil cooler; a, engine combustion air cooler; PU, pressurization unit.

a number of alternatives to consider relating to the type of heat generating plant for district heating – heating only, heating and CHP using gas, oil, coal or biomass, heating and CHP using refuse incineration. However, the CHP plant must have back up in the form of conventional boiler plant as shown in Figures 13.5, 13.6 and 13.7.

13.6   Accounting for linear pipe expansion

Whatever the piping material might be there is going to be movement in the distribution mains caused by the temperature rise and fall of the heating medium. There are various ways of accounting for thermal movement. What

Table 13.1   Coefficients of linear expansion for different pipe materials

Material	Coefficient of linear expansion m/mK (K^{-1})
Mild steel	12×10^{-6}
Copper	18×10^{6}
Polythene	140×10^{-6}
Polypropylene	140×10^{6}
PVC	75×10^{-6}

is important however is to ensure that the gradients to which the pipework is laid are not altered by its expansion and contraction. Otherwise the original venting and drain locations will not remain at fixed points in the pipe network. In other words, the thermal movement of the pipework must occur in two dimensions only, namely laterally and along its axis. Vertical movement will alter the gradient set for the pipe.

Table 13.1 gives coefficients of linear expansion for various piping materials. Using the coefficient for mild steel 100 m of pipe taken through a temperature rise from $10\,^{\circ}$C to $85\,^{\circ}$C will yield an axial expansion of $0.000012 \times 100 \times (85 - 75) =$ 0.09 m or 90 mm. If flow temperature is $140\,^{\circ}$C the axial expansion is 156 mm. One way to inhibit the effects of linear expansion is to pre-stress the pipe by heating it before jointing. This can be achieved by electrically heating the pipe. When different materials are used, as may be in the case of pipes within pipe, differential expansion has to be accounted for by the manufacturer of the external mains. Movement resulting from changes in temperature of the heat transfer fluid must be taken into consideration in the methods used for laying the distribution pipes.

One method is to lay the pipes in a preformed underground duct with expansion accommodated by natural changes in direction, this requires the use of pipe guides to ensure only lateral and axial movement and anchor points to control the direction of pipe movement. Figures 13.8, 13.9 and 13.10 show the site made or prefabricated expansion loop, double set and dog leg.

Another method employed to accommodate the linear pipe expansion is to use axial compensators as shown in Figure 13.11. These require pipe guides to control lateral and vertical pipe movement and anchors to control the direction of that movement.

13.7   Determination of anchor loads

It is important to ensure that forces present in the distribution mains as a result of thermal movement are held within the limits of the maximum stress and modulus of elasticity of the materials used. If the pipework is not subject to thermal pre-stressing during installation, a system of guides and anchors will need to control the effects of expansion that result from the rise in temperature of the heat transfer fluid. Pipework can also be pre-stressed-cold during

Figure 13.8 The prefabricated and site made expansion loop formed by natural changes in direction.

Figure 13.9 The prefabricated and site made double set using natural changes in direction.

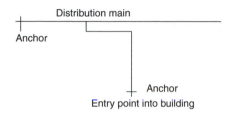

Figure 13.10 The prefabricated and site made dog leg.

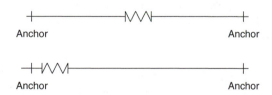

Figure 13.11 Two locations for the axial compensator between anchor points. Pipe guides omitted.

installation. This process is called cold draw. Having calculated the amount of expansion between anchor points, up to 50% can be taken up by forming a gap in the final joint and drawing the two pipe ends together for welding using a clamp. Thermal pre-stressing and cold draw puts the pipework system in tension when cold. Thermal pre-stressing and cold draw can only be employed where pipe expansion is accommodated by using natural changes in direction. It must not be employed when using axial compensators since they are designed to withstand compressive force not tensile force.

ANCHOR LOADS EMPLOYING AXIAL COMPENSATORS
[*Axial Compensators*, by N. Minkin and Sons, Ltd.]

There are two calculations necessary here. The first is to determine the anchor load under working temperature and pressure and the second is to check the anchor load when the pipe section is under test pressure. The higher of the two loads is used for designing the anchor housing. Determination of the anchor load under working temperature and pressure:

F_t = total force on the anchor (kN) = $F_1 + F_2 + F_3$

F_1 = force due to internal pressure on the bellows

= working pressure × effective area

F_2 = force to compress the axial compensator

= (force/mm) × mm of expansion

F_3 = the reactive force to that which overcomes the friction caused by pipe movement through the guides/ supports.

Clearly F_3 is dependent upon the coefficient of friction between the pipe and its guide or support.

For sliding joints a typical figure of 30 N/m of pipe is used for each 25 mm of pipe diameter. For pipe hangers and pipe rollers a typical figure of 15 N/m of pipe is used.

Example 13.1
Consider a 40 m length between anchor points of 150 mm bore carbon steel pipe under a working pressure of 400 kPa gauge and carrying water at 85 °C. Assume the pipe is guided in sliding supports and that the initial temperature is 10 °C. Select a suitable axial compensator and determine the net load on each anchor.

Solution
From Table 13.1 the coefficient of linear expansion for carbon steel is 12×10^{-6} m/mK. The amount of expansion $= 0.000012 \times 40(85 - 10) = 0.036$ m $= 36$ mm. From the Minikin catalogue a 150 mm bore type BAF Emflex flanged expansion joint is appropriate. It has a maximum axial compression of 50 mm that is well outside the actual amount of expansion. The force to compress the joint is 310 N/mm and its effective area is 252 cm².

$$F_1 = 400 \times 252 \times 10^{-4} = 10.08 \, \text{kN}$$

$$F_2 = \left(\frac{310}{1000}\right) \times 36 = 11.16 \, \text{kN}$$

$$F_3 = 30 \times 40 \left(\frac{150}{25}\right) \times 10^{-3} = 7.2 \, \text{kN}$$

$$F_t = 10.08 + 11.16 + 7.2 = 28.44 \, \text{kN}$$

If this force is expressed as a mass, the net load is $(28.44 \times 10^3)/9.81 = 2900\,\text{kg}$.

Summary

From Figure 13.11, you can see that there are two anchors supporting the compensator. As it is difficult to foresee how the anchor load will be distributed between each anchor the net load of 2900 kg is taken as applying to each one. You will appreciate that a net load of 2.9 tonne is substantial and serious consideration must be given to the structural design of the anchor housing.

If the test pressure is twice the working pressure then, $F_1 = 800 \times 252 \times 10^{-4} = 20.6\,\text{kN}$. This is less than the total force under operating conditions which clearly applies in this case in the calculation of the anchor load. It is worth noting the working pressure that the compensator is designed for and from the catalogue this is 1600 kPa. Had the test pressure been greater than this the compensator would need to be removed for the duration of the pressure test and a distance piece inserted in the pipe in its place.

ANCHOR LOADS EMPLOYING NATURAL CHANGES IN DIRECTION OF THE PIPE

The determination of anchor loads where the flexibility of the pipe is used to absorb thermal movement requires a knowledge of bending moments. Figures 13.11 and 13.12 show the deflection caused by a pipe moving axially as a result of thermal expansion between two anchors. These diagrams will be used to develop the formulae for offset (L), stress (f) and anchor force (F) that can be applied to the expansion loop, Figure 13.8, the double set, Figure 13.9 and the dog leg, Figure 13.10. Table 13.2 gives the modulus of elasticity (E) and the working stress (f) for carbon steel and copper.

From Figure 13.12 the deflection at the mid point of the offset L is $w/2$. From the bending moment diagram Figure 13.13, the mid point of the offset is $L/2$.

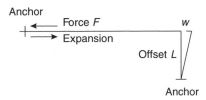

Figure 13.12 Linear expansion of pipe causing deflection of offset L by w metres.

Table 13.2   Young's modulus and maximum stress for mild steel and copper pipes

Material	Modulus of elasticity E Pa	Stress f Pa
Mild steel	200×10^9	60×10^6
Copper	70×10^9	60×10^6

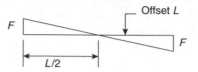

Figure 13.13   Bending moment diagram for offset L in Figure 13.12.

Thus

$$w/2 = F\left(\frac{(L/2)^3}{3EI}\right) = F\left(\frac{L^3/8}{3EI}\right)$$

from which

$$\text{force, } F = \frac{3EIw8}{2L^3}$$

and therefore

$$F = \frac{12EIw}{L^3}$$

From Figure 13.13 the bending moment in the offset is

$$F \times \frac{L}{2} = \frac{fI}{z}$$

from which

$$F = \frac{2fI}{zL} \quad \text{N}$$

Putting these two equations for F together,

$$\frac{12EIw}{L^3} = \frac{2fI}{zL}$$

then we have

$$\frac{2f}{z} = \frac{12Ew}{L^2}$$

and rearranging

$$L^2 = \frac{12Ewz}{2f}$$

Substituting

$$z = \frac{d}{2}$$

we have

$$L^2 = \frac{6Ewd}{2f}$$

For carbon steel pipes $E = 200 \times 10^9$ Pa; and working stress $f = 60 \times 10^6$ Pa. Substituting into the equation for L^2, we have

$$L^2 = \frac{6 \times 200 \times 10^9 \times w \times d}{2 \times 60 \times 10^6}$$

from which for carbon steel

$$L = 100\sqrt{wd} \quad \text{m}$$

Substituting for copper

$$L = 59\sqrt{wd} \quad \text{m}$$

You should now confirm that the formula for copper is correct.

Note that the working stress (f) is used in each formula and therefore this yields the minimum length for the offset (L). It will be useful to determine the stress (f) in the offset pipe when its length is greater than the minimum and since

$$L^2 = \frac{6Ewd}{2f}$$

then by rearranging

$$f = \frac{6Ewd}{2L^2} \quad \text{Pa}$$

Summary of formulae

$$F = \frac{12EIw}{L^3} \quad \text{N}$$

$$F = \frac{2fI}{zL} \quad \text{N} \qquad \text{where } z = \frac{d}{2}$$

$$L^2 = \frac{6Ewd}{2f}$$

Table 13.3 Moments of inertia for mild steel and copper pipes in cm^4

Pipe size (mm)	25	32	40	50	65	80	100	150
Mild steel heavyweight	4.29	9.16	13.98	30.8	64.5	114	272	862
Copper table X						24.4	71.4	304

For steel

$$L = 100\sqrt{wd} \quad \text{m}$$

and for copper

$$L = 59\sqrt{wd} \quad \text{m}$$

where offset (L) is a minimum

$$f = \frac{6Ewd}{2L^2} \quad \text{Pa}$$

Moments of inertia (I) for carbon steel and copper pipe are given in Table 13.3 in cm^4. The location of the anchors in Figure 13.13 has been used to obtain the formulae above. Figures 13.8, 13.9 and 13.10 show the practical locations for the anchor points in which, offset (L) includes two bends rather than one in Figure 13.13. For this reason the deflection (w) which is in fact the expansion resulting from thermal movement of the pipe can be halved so the deflection is $w/2$.

Example 13.2
Assuming that instead of the axial compensator used in Example 13.1, it is possible to reposition the pipe so that a double set can be employed between the two anchors 40 m apart, find the net load on each anchor.

Solution
Deflection w is 0.036 m. Since two bends are used in the offset, deflection is (0.036/2) m. For carbon steel minimum offset

$$L = 100\sqrt{wd} = 100\sqrt{\frac{0.036}{2} \times 0.15} = 5.2 \, \text{m}$$

$$z = \frac{d}{2} = \frac{0.15}{2} = 0.075 \, \text{m}$$

and therefore

$$F = \frac{2fI}{zL} = \frac{2 \times 60 \times 10^6 \times 862}{0.075 \times 5.2 \times 10^8} = 2652 \, \text{N}$$

Note the moment of inertia (I) from Table 13.3 has the units $862\,\text{cm}^4$ hence 10^{-8} is used to convert I to m^4. Force to overcome the resistance through similar guides to those used in Example 13.1,

$$F_3 = \frac{150}{25} \times 30 \times 40 = 7200\,\text{N}$$

$$\text{Total net force} = 2652 + 7200 = 9852\,\text{N}$$

Thus, net load on each anchor is $9852/9.81 = 1004\,\text{kg}$.

Summary
The net load on the anchor using a double set in the pipeline is $1004\,\text{kg}$. This compares with $2900\,\text{kg}$ when using the axial compensator. Clearly the use of natural changes in direction to absorb thermal pipe movement produces a significantly lower anchor load. Natural changes in direction for the distribution mains is therefore the better option in accounting for pipe expansion. Furthermore pipe and pipe fittings are mechanically very strong whereas the axial compensator introduces a weak point in the pipe section.

Example 13.3
A 100 mm bore steel pipe is located between two anchor points to accommodate linear pipe movement as shown in Figure 13.12. Its deflection due to linear expansion is 75 mm. Taking the coefficient of friction for roller pipe guides as 0.3, determine:
  (i)   the minimum length of the offset and the net force on the anchors;
 (ii) the length of the offset if the stress (f) in the pipe is to be limited to 50% of its working value;
(iii) the revised net force on the anchors.

Solution
Note: you will appreciate that the offset (L) in Figure 13.12 will also expand axially.
  (i) Adopting the formula using working stress,

$$L = 100\sqrt{wd} = 100\sqrt{0.075 \times 0.10}$$

from which the minimum length of the offset (L) is $8.66\,\text{m}$.
    Net force on the anchor

$$F = \frac{2fI}{zL} \quad \text{N}$$

where

$$z = \frac{d}{2}$$

$$F = \frac{2 \times 60 \times 10^6 \times 272}{0.05 \times 8.66 \times 10^8} = 754 \, \text{N}$$

Net force on the anchors including the resistance through the guides

$$F = 754 + (754 \times 0.3) = 980 \, \text{N}$$

(ii)   Adopting the formula

$$L^2 = \frac{6Ewd}{2f}$$

where

$$f = (30 \times 10^6) \, \text{Pa}$$

$$L^2 = \frac{6 \times 200 \times 10^9 \times 0.075 \times 0.10}{2 \times 30 \times 10^6} = 150$$

So offset (L) is 12.25 m.
   (iii)   Adopting the formula

$$F = \frac{2fI}{zL} \quad \text{N}$$

where

$$z = \frac{d}{2}$$

Net force on the anchor

$$= \frac{2 \times 30 \times 10^6 \times 272}{0.05 \times 12.25 \times 10^8} = 266 \, \text{N}$$

Net force on the anchors including the resistance through the guides

$$F = 266 + (266 \times 0.3) = 346 \, \text{N}$$

Summary
You will notice that whilst one of the anchors in Figure 13.12 is subject to a compressive force the other is subject to a bending force or turning moment. In solution (ii) the net compressive force is 980 N and the net bending force would be taken as 980 N. The anchor housings would need to be designed accordingly. There is a significant reduction in the force on the anchors when the working pipe stress is halved in part (iii).

Example 13.4

A site steam main layout is shown in Figure 13.14. Anchor A1 is at the exit to the plant room and anchors A4 and A5 are at the entrances to buildings. These anchor locations ensure that forces generated by pipe expansion are not transferred to pipework within the buildings. Working steam pressure is 400 kPa gauge and initial temperature can be taken as 10 °C. The coefficient of friction for the guides serving the axial compensator is 15 N/m of pipe for each 25 mm of pipe diameter. The remaining guides have coefficient of friction of 0.4.

Given that the steam main is mild steel, identify which anchor has a differential force applied to it and determine the net forces imposed on the anchors.

Solution

From the *Thermodynamic and Transport Properties of Fluids* (by Rogers and Mayhew) steam temperature at 500 kPa absolute is 151.8 °C.

(i) *Anchors A1–A2*

$$\text{deflection } w = 0.000012 \times 25 \times (151.8 - 10) = 0.0425\,\text{m}$$

From the Minikin & Sons brochure an Emflex type BAF 100 mm bore axial compensator can be compressed to 50 mm, so 42.5 mm of expansion is within its capabilities. The force to compress the compensator is 80 N/mm and its effective area is 121 cm².

$$F_1 = 400 \times 121 \times 10^{-4} = 4.84\,\text{kN}$$
$$F_2 = 80 \times 10^{-3} \times 42.5 = 3.4\,\text{kN}$$

and

$$F_3 = 15 \times 25 \times 100/25 \times 10^{-3} = 1.5\,\text{kN}$$

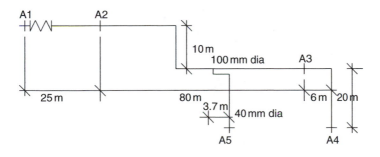

Figure 13.14   Site steam main layout for example 13.4.

So the total net force is $4.84 + 3.4 + 1.5 = 9.74\,\text{kN} = 9740\,\text{N}$. This force is assumed to apply to both A1 and A2.

(ii)   *Anchors A2–A3*

deflection $w = 0.000012 \times 80 \times 141.8 = 0.136\,\text{m}$ expansion

$$f = \frac{6Ewd}{2L^2}$$

As there are two bends in the double set $w/2 = 0.068\,\text{m}$. Thus

$$f = \frac{6 \times 200 \times 10^9 \times 0.068 \times 0.1}{2 \times 10^2} = 40.8 \times 10^6\,\text{Pa}$$

This is within the working stress for mild steel given in Table 13.2.

$$F = \frac{2fI}{zL}$$

where

$$z = \frac{d}{2}$$

and substituting

$$F = \frac{2 \times 40.8 \times 10^6 \times 272}{0.05 \times 10 \times 10^8} = 444\,\text{N}$$

Note the term 10^8 which converts the moment of inertia (I) from Table 13.3 to m^4. Accounting for the pipe guides, net force (F) is $444 + (444 \times 0.4) = 621\,\text{N}$. This force is assumed to apply to both A2 and A3. The differential net force on A2 is $9740 - 621 = 9119\,\text{N}$.

(iii)   *Anchors A3–A4*

deflection $w = 0.000\,012 \times 20 \times 141.8 = 0.034\,\text{m}$

$$f = \frac{6Ewd}{2L^2}\quad\text{Pa}$$

and substituting

$$f = \frac{6 \times 200 \times 10^9 \times 0.034 \times 0.1}{2 \times 6^2} = 56.67 \times 10^6\,\text{Pa}$$

This is below the working stress of $60 \times 10^6\,\text{Pa}$

$$F = \frac{2fI}{zL}$$

where

$$z = \frac{d}{2}$$

and substituting

$$F = \frac{56.67 \times 10^6 \times 272}{0.05 \times 6 \times 10^8} = 1028\,\text{N}$$

Using the alternative formula

$$F = \frac{12EIw}{L^3}$$

Substituting

$$F = \frac{12 \times 200 \times 10^9 \times 272 \times 0.034}{6^3 \times 10^8} = 1028\,\text{N}$$

Accounting for the pipe guides net force (F) is $1028 + (1028 \times 0.4) = 1439\,\text{N}$. This force is assumed to apply to both A3 and A4.

(iv) *Anchor A5* Deflection due to expansion over 20 m with two bends at the offset $= w/2 = 0.034/2 = 0.017\,\text{m}$.

$$f = \frac{6Ewd}{2L^2}$$

and substituting

$$f = \frac{6 \times 200 \times 10^9 \times 0.017 \times 0.04}{2(3.7)^2} = 29.8 \times 10^6\,\text{Pa}$$

This is within the working value of $60 \times 10^6\,\text{Pa}$.

$$F = \frac{2fI}{zL}$$

where

$$z = \frac{d}{2}$$

and substituting

$$F = \frac{2 \times 29.8 \times 10^6 \times 13.98}{0.02 \times 3.7 \times 10^8} = 112.6\,\text{N}$$

Accounting for the pipe guides net force (F) is $112.6 + (112.6 \times 0.4) = 158\,\text{N}$.

> Note that this anchor force is much less than that for anchor A4 due to it being a smaller size pipe and the offset has two bends.
>
> **Summary of net anchor forces**
> A1 $= 9740$ N, differential net force on A2 $= 9119$ N, A3 $= 621$ N, A3 is also subject to a bending force of 1439 N, A4 $= 1439$ N and A5 $= 158$ N.

13.8   Pipe guides and anchors

Details of pipe guides and anchors are found in the Applications brochure of the manufacturer of the axial compensator. A golden rule is to use material of the same dimensions as the pipe. Thus if a 100 mm bore mild steel pipe is being guided, supported or anchored, the channel iron or angle iron used needs to be 100 mm in section.

13.9   Further reading

The District Heating Association
Combined Heat and Power Association

13.10   Chapter closure

You have now completed the chapter on district and community heating and gained knowledge of its main features including factors for consideration, choice of system, distribution pipework, generating plant, accounting for linear pipe expansion and determination of anchor loads. This should provide you with a foundation in this area of specialist work.

Index

Page numbers in **bold** indicate figures, and numbers in *italic* tables